The *Definitive* Beginner's Guide to **Tendering & EU Procurement**

Procuring Contracts in Social Housing or other Public Body Environments

SECOND EDITION
Revised & updated to include the new 2015 Regulations

Andrew Shorter

Printed and bound in the United Kingdom by
4Edge Ltd, 7a Eldon Way Industrial Estate,
Hockley, Essex, SS5 4AD

Contents

List of Figures

*Yet again I dedicate this – the updated version –
to my family, where I hope my persistence will
eventually convince them that procurement is a
real thing.*

About the author

Andrew Shorter has worked in construction and procurement for over 30 years, primarily in a Social Housing environment. Working with Local Authorities, ALMOs and Housing Associations, his experience in this sector includes all aspects of EU and non-EU procurement and contract management. Andrew has worked as a procurement adviser, practitioner and trainer and continues to do so.

Preface

The purpose of this book is to provide advice and guidance on procurement for those in Local Authorities and other public-type bodies who have been charged with tendering building repair, maintenance or construction contracts. Written primarily with housing and other property professionals in mind, it assumes only a rudimentary knowledge of procurement and so serves as an introduction to the craft for the less- or the completely inexperienced. Nevertheless, it also assumes that the reader will need to take procurement exercises through to their completion and in this vein attempts to cover – in plain language – the whole of the process from conception through to the start of the awarded contract. In this respect, it is hoped it may also be of some use to a wider or more experienced audience.

Health warning: this is not a legal textbook. It has, in this second edition, been updated to take into account the new 2015 Public Contract Regulations and will cite Laws and Regulations, but it does not cover all the new Regulations *per se* and is not intended to replace any legal guidance, whether written or obtained from a lawyer or other appropriate authority in the flesh. It will, however, advise you on the relevant and key pieces of legislation that you need to be aware of, to work within and probably to seek more detailed (and proper) advice on[1] as you work through the book.

With additional content, the aim has once again been to create a book that is at once readable (on procurement?) yet sound in advice and somewhat broader in content than the first edition. It has been written envisaging that the reader might well be reading the chapters marginally ahead of the process, so the contents attempt to generally follow a logical order, going step by step through all the necessary stages, but the nature of procurement means that this rule has been unavoidably transgressed in some areas.

---

1. I apologise for ending a sentence on a preposition. Also – always ensure you remain updated on laws and regulations – all books giving references become out of date eventually, if not rapidly.

Sometimes, a particular stage in the process involves many aspects that could, in reality, populate an entire book on their own, so some sections may seem to digress widely off the mark. This level of coverage is intentional, as I have tried to provide advice and guidance across the *whole* procurement process within the chosen field. In similar vein, some duplication of topics will be apparent.

On the other hand, there are many nuances in procurement that one would benefit from knowing but can only acquire with time and experience. I can offer you tantalising tasters such as "Stadt Halle and the Teckall Exemption" (no – not a new Harry Potter movie), "Beentjes" and "Buglife v. Medway Council": these are but a few of many, many cases where landmark decisions have been made about EU procurement law that have necessitated a tweak in the approach you take to a tendering exercise. There are too many such examples to list here, and if I did you would probably not read them[2], so I have stuck to the firm and basic and leave it to you to seek a broadening of your peripheral procurement knowledge. How you can do this, however, *is* covered later on.

As a bonus, it may be found that this book is of practical help to those studying procurement in a broader or more formal setting. To this audience I offer a muted apology that the book is somewhat restricted to the theme of works and repairs. A true student of procurement will need to cover a much wider range than this but the fact is that a book covering the whole gamut of procurement would either have to be very general, five times as thick or include copious amounts of clarifications and conditions. It was felt by me that such an approach would only serve to distract the reader when the topic I have covered is gargantuan enough.

The main thrust of this book is aimed at the tendering of building maintenance and repairs contracts and construction projects but it also has property service contracts (such as grounds maintenance and cleaning) in mind as well. However, please don't be put off if your procurement 'bag' does not lie in any of these domains: many

---

2. See my other title *EU Procurement: Legal Precedents and their Impact*, published by Cambridge Academic in 2014, for examples!

aspects of procurement are common to all areas of the profession so I hope the scope covered is such that students and professionals in other spheres of work may also find it of help.

It will be seen that I cover carrying out tender processes in the 'old fashioned' way, using e-mail and hard-copy documents, but time, and the EU Regulations, move on and e-tendering portals are becoming the norm. For this reason, throughout most of the book, I assume tendering will be carried out in this way, taking advantage of the benefits it can afford.

I have tried to make this tome at once enjoyable and informative and hope the adopted style suits you. I am a practitioner, not an academic or theoretician: this book is therefore a practical guide and not a book of theory, and I make no apologies for that.

In conclusion, I hope you enjoy reading this book more than I have enjoyed (re-)writing it and I hope it helps you towards good procurement practice.

Andrew Shorter

Introduction

I have often led seminars and training sessions where I open with the question: "What is Procurement?" The stock answer that comes back is "buying" and I have to counter that it isn't – not quite, anyway. Dictionaries will often define procurement as buying, and the link is there, but for us, in local authorities and similar organisations, buying is only a *part* of the art of procurement – procurement, as we have to do it, is different.

People who are "buyers" respond to job adverts for buyers: supermarkets and car manufacturers use people called 'buyers' to source products to go on their shelves because they actually do go out and buy. Buyers *buy* things and generally buying involves a relatively simple balance comprising mainly cost and quality[3]. But procurement – especially public body procurement - is not just that: it is so much more.

"Procurement" in the context of this book (and which you will understand if you are reading it) is an altogether more complex process that covers a whole life cycle from initial concept and definition of business need through acquisition of the 'product' to the end of its life; it encompasses a range of skills and disciplines working together over a period of time to deliver under best value for money criteria anything from a box of staples to a large, complex or innovative contractual commitment, all within the confines of changing international law and internal regulatory frameworks. It really is not simply going out and 'buying' something.

This book is not about buying, but about procurement in its fullest sense, and the procurement of works and repairs contracts in particular – and you cannot simply go out and 'buy' these types of contract.

In practice, procurement comprises the obtaining of goods or services using a cornucopia of tendering, specifying, complying,

---

3. Not quite true – buyers will rightly point out that many more elements form part of the decision equation, including availability, delivery, warranties, and so on. Nevertheless, buyers by definition are less restricted than procurement folk in the way they are able to secure the goods.

evaluating, assessing, legalling[4] and lots, lots more: in other words the combination of a whole mass of skills, knowledge and artistry that would baffle many lesser people, and this book intends to open up this Pandora's Box of goodies and explain it all to you in simple language.

In the interests of completeness, I will touch on the word 'Commissioning', often used in place of the term 'Procurement' and the cause of great debate. The difference – if indeed there is one – is miniscule. The Audit Commission defines Commissioning as "the process of specifying, securing and monitoring services to meet people's needs at a strategic level"[5] - in other words defining the need and planning the solution. Some people disagree with that and say 'Commissioning' is actually securing the 'product' once the contract for its supply has been established (through a procurement, obviously). Take your pick. Procurement in this context is seen as the practical process of actually securing the provision. I am henceforth going to ignore this debate and what follows applies to Commissioning as well as Procurement. For the purposes of this book, and to keep me happy, just avoid "buying"!

Regardless of your job title, if you are in a Local Authority or similar public body and you have been asked to tender a contract, you will be involved in a procurement process. The purpose of this book is to explain to you step by step how to go about this in a way that will achieve the key aims for any procurement exercise, namely:

i) To obtain a service of a quality that meets, at least, your minimum requirements (or your client's[6]).

ii) To obtain it at the best price you can.

iii) To secure the service under circumstances and conditions that are ultimately to the benefit of the client (i.e. you or those you serve) and safeguard their interests.

---

4. Legalling – made-up word for being unavoidably involved in a mass of EU and other legislation.

5. Social Services Inspectorate / Audit Commission (2003) *Making Ends Meet*

6. 'Client' may be an internal department. Who, effectively, commissioned you to carry out the procurement?

iv) To secure the service in a manner that is wholly compliant with the myriad of laws and regulations that apply to public-body procurement processes.

What does this all mean?

In the 'good old days', Local Authorities were driven by the government of the day to focus on cost alone and, always, cheapest was best. (I am not going to give a potted history of local government but bear with me – this is all relevant). This culminated in the era of Compulsory Competitive Tendering (CCT)[7] in the 1980s where *everything*, even the housing and other local authority services themselves, were provided by organisations who had won the contract on the back of a lowest-cost tender. There then came a revolution and it was realised that cheapest was not always best – newly-empowered 'customers' were beginning to question the quality of the services they were receiving and then came the concept of *'value for money'*. Value for money continues to this day and, abbreviated to vfm, it is fast becoming the procurement mantra of our time and, to be fair, the plain-English logic of the phrase cannot be contended.

I have, though, skipped a bit. Value for Money was preceded by another process known as Best Value (BV), which prescribed the route through which vfm (for in reality that is what it was) might be obtained. The older amongst us might recall the 'Four Cs of Best Value' – Challenge, Compare, Consult, Compete – and I do not see a lot wrong with that principle even today, despite the fact that Best Value as it was has now gone out of official 'procurement' fashion[8].

I will, in fact, be drawing on some of these 'old' BV principles from time to time as we progress through the book so I shall briefly explain the Four Cs so as to put the uninitiated amongst you out of your misery:

- **C**hallenge – always challenge all aspects of an existing contract or service at time of renewal – never assume any

---

7. A Conservative Government was in power at the time
8. The term is still used today, however, but in another context. Guidance was issued on this in September 2011 and I refer you to Chapter 16.

of it is right or good or basically OK to leave alone. I always say – 'Ask "why?"'. I keep doing this and people often find it annoying, but it is surprising (and amusing) how often people don't know the answer. If they don't know the reason, why are they doing it[9]?

- **C**ompare – make comparisons with other similar client bodies' contracts of similar nature[10] to see how you are doing compared with them and, if they have good ideas, pinch them. Then you can donate your good ideas to them. That is procurement good practice – pinch and share. Remember that.
- **C**onsult – always seek input from 'stakeholders' – in most cases, these are users and experts who between them have a pretty good knowledge of the customers' needs and the best ways to meet them.
- **C**ompete – this is the actual tendering process to get what you require at the best possible price (note – not necessarily the cheapest price. We are, after all, after best value).

So 'Best Value' as it was is no longer with us and new methods and terminology (and therefore jargon) have taken its place, but the principles that it expounded were sound at the time and remain so today. Procurement people are not, by their very nature, dinosaurs and are always keen to find and exploit the new – that is, after all, their job – but we lose the good stuff[11] from the past at our peril.

Reverting back to my three aims of procurement, I shall elaborate on these, now, as the attainment of these objectives forms the core *raison d'être* of this book. It is why we're here.

Primarily, all procurement is aimed at meeting a need to the best standard possible. This is why the consultation element is so important. How can you meet a need if you don't know what that need is?

9. This principle rears its head when we cover evaluation, too. Keep reading!
10. Comparison is often made through Benchmarking and Benchmarking Clubs are good for this.
11. Good stuff – good practice that stands the test of time (my definition).

Securing the need rarely stands as the sole criterion for measuring the success of a procurement – seldom is cost not a factor. It may seem sometimes that Government procurement disregards cost (and maybe in some instances it does), but Government procurement is a bad example of the art and I shall always assume that cost is a factor, albeit lowest cost is not always the key consideration. Just remember – you will always be spending other peoples' money (often, in the end, the money of those who pay your wages) so you will always have the responsibilities associated with being the holder of a public purse.

Finally, circumstances and conditions. There will always be conditions attached to paying someone to provide a service and deciding on these conditions is part of the process. Some conditions – laid down in the contract – are common to many procurements (with caveats, as we shall see) and some conditions are determined by you to make the service you have secured more specific to your particular needs and – very importantly – its delivery more manageable once the contract has started.

You will note I have used the word service a lot, but beware - I have used it to mean anything you get from a provider. In the context of this book it could be the construction of a building or it could be a boiler repair service (which would be a service in the true sense of the word). In the world of procurement – as in its definition, above – 'Services' is something that is neither works nor goods, and this is a particularly important delineation when discussing any procurement that comes under the realms of European (EU[12]) Law (see later).

You will need to bear with me on this. Where I refer to Services in the context of EU procurement, I will make it plain and capitalise the 'S'. Where I am using it as a generic term for what we get from contractors in the way of works or repairs, I will use a small 's'. This distinction may at the moment seem a little convoluted and pedantic, but use it I will, as repeated use of the phrase 'works or

---

12. EU – means the European Union, or more frequently for us, in the context of "the EU Regs" or "the EU Laws", used familiarly to refer to the European Procurement Regulations.

repairs' is cumbersome and unnecessary. Things may, though, get tricky where the word 'service' comes at the beginning of a sentence.

Interpreting the Regulations

In the four bullets above I referred to the importance of compliance and throughout this book I make much of it, and, believe me, rightly so but I also need to make a strong point about context.

You cannot transgress UK or EU law and you cannot carry out a process in contravention of your organisation's internal regulations. However, the way certain elements of law are interpreted and the restrictiveness of an organisation's internal regulations will vary from one organisation to another.

In general, you will find that Local Authorities tend to be very risk averse and much more compliant and cautious in the way they interpret the rules, and this is understandable – they have a duty of care, responsibility for a public purse and an onus put upon them to be bastions of good practice. Local Authorities will also have multiple layers of bureaucracy to ensure these obligations are met and no commercial pressures to encourage any deviation.

On the other hand, in the world of non-local authority social housing, such as a Housing Association, the approach is often quite different. Because they are spending public money (from rents and government grants, for example), they have to observe the EU rules but they are much more commercially-driven and much less bound by (local) government red tape. For this reason, you will find their interpretation of compliance with the rules often seems to be much more – um – lax. This does not mean they do not comply, or that they take undue risks, but they do have commercial pressures and with these comes more (commercial) freedom to be less demanding in the way they interpret the regulations.

In between, you have ALMOs[13], and their approach to procurement rules and regulations will be somewhere in between as well.

---

13. Arm's Length Management Organisations that have been set up by a Local Authority to manage their properties, so are almost independent but still come under the Local Authority wing so are not completely independent.

On this question of context, I have no strict guidance for you – it is entirely up to the organisation itself how far they go in relaxing their demands for overt compliance (their risk, if you like) and you will need to clarify this as you move from one position to another. If you think something is too lax or too strict, you need to ask and you may then be assured about the proposed action or you may have raised a valid point and saved someone a whole lot of problems.

For my part here, I am showing you the cautious, compliant, Local Authority-type approach that will always be lower-risk. That is because I wish you to be safe from challenge and less likely to get the sack. If you are with an organisation that has a more relaxed approach than this book would recommend, then read the previous paragraph again and make your choice.

Just remember – in procurement only a few questions have a yes/no answer; the vast majority all have an answer that begins with "That depends..." Questions regarding compliance come in the second category.

Enjoy it!

We are nearly there, now – that is, the part of the book where I actually start imparting guidance and – I hope – some wisdom. Before I do, though, I have just one final thing for you to bear in mind that is core to the whole procurement scene. This book will concentrate on works and repairs contracts procurement, but an awful lot of what it tells you will be transferable to the procurement of other services, such as the supply of goods and Services (yes - note the big 'S').

The point is this – when you procure, the door is open to avenues of experience to which you never thought you would or could ever aspire, and through this you will discover a lifetime of learning and discovery and a path that leads onto a raft of opportunities and copious metaphors.

To make this journey, you do not need to know everything about everything, but you will be surprised how much you learn. You need to know about procurement, and then you need to know how to get on board those who do know about the subject you are working on. That is all.

This book will set you firmly on this path. I wish you well.

1 // The Gentle Art of Procurement

Yes, the title is a little tongue-in-cheek, but the aim of this first chapter is to give you a *feel* for the procurement process – the 'arty' bit – before we come to the grittier, specific aspects where the 'science' kicks in.

Procurement is bound up in various layers of regulation and you need to be careful to adhere to all of them! Experience will build within you a bank of knowledge that will assist you with compliance but this can never fully replace the on-hand knowledge of a legal officer on your project panel. The regulatory frameworks all deal with process, and it is process that forms the basis of (my figure) 90% of all legal challenges[1] to a procurement exercise. Do good process, and you will hopefully avoid too many challenges. You will never avoid all of them.

Internal Regulatory Frameworks

The internal laws of an organisation normally comprise Contract Standing Orders (CSOs) or their equivalent[2] and Financial Regulations. You may also have a Procurement Code of Practice specifying *how* the CSOs are to be applied. Be warned: in most organisations, contravention of these internal rules can lead to dismissal. If procurement hits a major problem, but you have complied with your organisation's internal regulations, you will generally be OK. If you have not – consider packing.

UK Statutory Law

Until recently, UK Law in general applied to the contracts themselves rather than the tendering stage, but the Remedies Directive of 2009 (and now the Public Contract Regulations of 2015) brought powers under European Procurement Law into the UK courts. These powers are fairly draconian and will be looked at later, but they leave you the procurer much more vulnerable in the event of a challenge than you ever were before.

1. Challenge – See Glossary
2. For example a Procurement Code

European Law

European Procurement Law is the defining law of procurement process. The EU Law is translated into English Law as the (now) Public Contracts Regulations 2015 (called the 'EU Regs') which can be viewed online if you wish[3]. This is not a book on law, so we will not look at these Regulations in any great or specific detail, but we will draw on them as required and refer to them enough to make sure you know which furrow to plough, and when to seek detailed legal advice.

The EU Regs 'kick in' when a contract being tendered exceeds a certain value or threshold. Known as the EU Thresholds, the values for Works, Supplies and Services (capital 'S' – see 'Introduction') are determined in January every two years. The last time the figure was set was January 2016, when the works threshold, for example, was set at £4,104,394 and Supplies and Services£164,176. The numbers are odd because they are laid down in Euros and converted into Pounds Sterling.

I refer to Works, Supplies and Services, but there are also thresholds that apply to Utilities and other classifications of contracts. These types of contract are outside the remit of this book but the procurement processes vary very little, so utilities procurers will still hopefully find the advice contained in this book to be of help.

Whilst the EU Regs apply in full for projects above the threshold, you need to be aware that they still have influence on procurements below that value: under the terms of the EU Treaty, *all* procurements have to be carried out in a way that is non-discriminatory, transparent, proportionate and competitive.

EU Law overrides all internal regulations, and most internal regulations (such as CSOs) state that all procurements must be conducted in line with EU Law and where EU Law contradicts the Internal Regulations, EU Law shall prevail.

Be advised – don't flout any of them.

---

3. Available online at http://www.legislation.gov.uk

The 'new' Public Contracts Regulations 2015 – a note

A special note on these Regulations because there have been many changes made to certain aspects of the more familiar 2006 Regulations and some parts of them are much more complex.

I have stated that this is not a book on law, and I stick to that premise. I have therefore not gone into detail on all of the new Regulations or dwelt on their subtleties (other books can do that), but have dealt with those that are relevant to the topic. To understand all the Regulations you will need to do three things:

1. Read the Regulations – you need to go to the following web address and download them:

    http://www.legislation.gov.uk/uksi/2015/102/pdfs/ uksi_20150102_en.pdf

2. Read the Guidance issued periodically by the Crown Commissioning Service. This can be downloaded at:

    https://www.gov.uk/transposing-eu-procurement-directives

3. You are then strongly advised to refer regularly to the Procurement Policy Notes issued from time to time by the Cabinet Office at:

    https://www.gov.uk/government/collections/procurement-policy-notes

The reason I advocate this approach rather than iterate the various guidances here is because the guidance is constantly changing: new Regulations require interpretation and testing and the 2015 Regulations still (almost) have the wrapping on them so there is (at the time of going to print) a lot of uncertainty about what the

Regulations actually mean and how apparent contradictions will be resolved. Their interpretation in the courts will be the next step...

For this reason, in this second edition, I may caveat some of what I advise with a caution based on this degree of uncertainty and I apologise for that.

Procurement Routes and Processes

All procurements follow one of a range of processes known as routes and, regardless of the chosen route, they all follow a core path made up of component parts or stages. Sometimes these component parts are sequential and sometimes they can be run in parallel, and working out what has to be done when is the art of timetabling. Remember – the key to a successful procurement is the timetable. The list of tasks that need to be in a timetable is extensive and will vary in detail according to [a] the type of route you have selected and [b] the nature of the organisation in which you operate. Nevertheless, like any toolkit, you can reckon the component parts you will need in most, if not all, procurements will include:

- Planning, strategy and timetable
- Market research and engagement
- Approval to proceed
- Preparing for a bidder selection process
- Writing specifications
- Collating data and information
- Consulting with stakeholders
- Drawing up contracts
- Advertising
- Selecting bidders
- Preparing tender documents
- Tendering
- Evaluating tenders
- Seeking authorisation to award
- Sitting out Call-in, Scrutiny or Standstill periods

- Initialising the contract – the 'start on site'.

What is most interesting about this list is the fact that actual 'tendering' is only one, seemingly minor, part of the whole process. This is right: it is the preparatory work you put into it 'behind the scenes' that makes the exercise a success, and all of this constitutes 'procurement'.

'Tendering' *per* se can be undertaken in a variety of ways and we shall, in the course of this book, look at all of these ways (or routes) in some detail and we will certainly look in great detail at the two *key* procurement or tendering routes which are the basis of most procurement processes – the 'Open' and the 'Restricted' Procedures. These two between them could easily cover most of the procurements you will ever need to do.

There is another process, however, and for completeness we shall deal with it now and get it out of the way – 'seeking Quotations'.

Seeking Quotations

Seeking a quotation – or 'getting a quote' – is exactly that: seeking a simple price for a simple supply item. Used for items where the supply is concise and clear, the process does not provide for any comparisons between providers other than price so is only really good for the purchase of 'things' like refrigerators or bus hire for a trip. Your specification can be exact – model number, delivery time, installation – and price (provided all your potential suppliers are recognised as legitimate) becomes the only comparator.

Normally, goods or services supplied under a quotation process are done so on the supplier's terms and conditions – just as if you were buying in a shop. This is sometimes a drawback and can mean a quotation process is not appropriate and should not be used. Additionally, the simplicity of the process means that it is not suitable for purchases that are in any way complex. Horses for courses.

The number of quotations you have to obtain will be written into your organisation's Standing Orders (or equivalent), but will probably be around three or four. The only other proviso will be a cap on the value up to which you can simply seek quotations for a supply. This may be £25,000 or £150,000. If you are procuring, you will need to know this. Once your purchase value exceeds this limit, you will need to tender.

All this having been said, there are some variations on the theme: some consultancy services are straightforward enough to secure through a quotation exercise, you could, if you choose, go out for quotations using your own terms and conditions and you could, if necessary, include some quality aspects in your process, only then it would become what is, ostensibly, a mini-tender.

Tendering takes you into a world that can be much more complex than simply seeking quotations and that is the world where this book will dwell. Consider every step you take and your procurement will become a successful project, with all the satisfaction that brings.

Summary

- Procurement of any sort at any level is governed by internal and external regulatory frameworks.
- A procurement is a project with many distinct stages – each stage has to be considered carefully
- The two key tendering routes are the Open and the Restricted procedures
- Quotations can be sought for the purchase of simple, easily-defined items or services
- Planning and timetabling are key to a successful procurement.

2 // What (or who) are you working with?

At the start of this book there is a Preface and an Introduction, which I hope have given you an idea of how this book is designed to 'work'. Chapter 1 gave a broad insight into what a properly executed procurement comprises. I shall now give a brief outline of how maintenance, repairs and construction project contracts themselves work so that if you are new to this field you will understand the context of the procurement process and what it is trying to achieve. I would prefer it that anyone procuring a construction contract knows what a construction contract is, but I did say this book was for beginners, so I shall be true to my word. I also said a procurer can procure anything, provided they have the right skills around them. This is also true, but a little background knowledge never hurt anyone...

Repairs

In construction, and social housing construction in particular, there are principally two types of contract – repairs and project. The first type is normally set up to provide an ongoing ('day to day'), responsive repair service to maintain whole buildings, individual properties, heating systems, and so on. These are tendered and let for a specified time period and a new contract has to be procured and in place ready to start at exactly the time the current one finishes. Such contracts are normally rooted in a social landlord's legal obligation to maintain properties in a habitable condition and any lapse in this service provision must be avoided at all costs. These contracts are normally set up as *Term Contracts* (i.e. let for a specified term or period) and funded through revenue budgets.

Term contracts are paid on a monthly basis, with work generally priced against a tendered *Schedule of Rates*, that is, prices the contractor has set against a list of all the types of job it is envisaged the contractor will need to carry out during the course of the contract, and yes – it *is* a big list of jobs.

Projects

Projects are normally funded by capital (at least in part) and are tendered to achieve a specific, one-off goal that has a clear and finite end. The project may be building houses, refurbishing a block of flats or installing 275 new boilers; once the job is done, the contract finishes – sort of. It is normal in such contracts for there to be a *retention* of part of the costs (normally about 5%) throughout the project; when the job is complete, half of this is paid to the contractor and the remaining 2½% held on to for a specified time (often a year – depending on the type of work) called the *Defects Liability Period* (DLP). During this time, the contractor remains liable for any problems with the work and has to put them right before the final monies are paid over. This is in addition to any other form of guarantee that may be required by the client.

Projects are paid against *evaluation certificates*, which are issued by a Quantity Surveyor (QS) at times specified in the contract. These times could be monthly, or at distinct milestone stages in the project and the payments have the retention deducted. Term contracts do not always hold retentions as the client always owes the contractor money due to payments being made in arrears and there is always work in progress – there is no need to financially maintain the contractor's interest!

Quantity Surveyors (QS)

The Quantity Surveyor (nowadays often called a 'Cost Consultant') deals exclusively in quantities, numbers or amounts – length, area, volume and amount (amount of money, that is). They will be able to help you draw up the tender documents, prepare pre-tender estimates (what it is thought the contract will cost), set up methods to evaluate the bids to secure the best value, measure volumes and areas and quantities of materials, etc. Once the contract is let and functioning, a QS – maybe not the same one that helped to tender it, but it is possible (and often preferable) – will then work for the client on the contract's finances, working out what the contractor is owed, auditing invoices, measuring volumes of work completed, etc.

Building Surveyor (BS)

Quantity Surveyors are not *building* technicians, they are *number* people. If you want a building technician, you need a Building Surveyor. They are experts on the technical (i.e. construction and regulatory) side of building and will help you specify what the contract requires and can help you write it into the specification (or even do it for you, if you pay). When the contract is running they will help define what work is required on site, monitor what work has been done and identify what is ready and OK for payment. The QS will then value this work.

Architects and Engineers

Surveyors can tell you what work is required, but they do not design. Architects design buildings, and engineers design plumbing, electrical, air conditioning and heating systems. On some contracts you tender, you may require the services of both these disciplines.

On most contracts you will certainly need a QS and on many you will require a Surveyor. You need to construct your procurement team carefully and make sure you know what skills you will require. It may be that you do not have all the required skills to hand within your organisation and if so, before you can tender the main contract, you will have to tender for these missing skills and appoint consultants to provide them. We will show you how this is done, so fear not – just remember to add on to your project the additional time required to do it.

Contract Manager

One more point on procurement teams. I will labour the need to build into a contract's documentation the tools to manage it. For this reason, you should have in your team the Contract Manager or Administrator who will be running the contract once it is let. Hopefully, they will be practiced in this role and be able to provide help and guidance on what tools to include. Even if they are not, if they will be running your contract you want them on board so that they have input and therefore *ownership*. In this way you will get continuity from the procurement stage to the operation. Your contract will be run as you envisaged.

This, in a nutshell, is where the different skills fit into your procurement world. We will now look at procurement itself.

Summary

- A Quantity Surveyor works on numbers and money – quantities and prices
- A Surveyor is the expert on building technology and specifications
- Architects design (buildings and other things)
- Engineers design a building's services and systems
- Contract Managers make the contract happen
- Term contracts provide for ongoing maintenance and repair
- Project contracts result in a tangible asset or improvement that generally add value
- You will need some or all of these skills in your procurement team, depending on the nature of the contract.

3 // Procurement Routes

What is an 'EU Procurement'?

I need to clarify what is meant by EU and non-EU procurement and explain how we will deal here with each. The European Union (the EU) has laid down specific requirements on the way tendering should be carried out in Europe (i.e. by its Member states) to ensure that the process is fair and non-discriminatory to all companies across the whole of Europe, is transparent in its approach and is competitive. These conditions are ensured through European Legislation or Directives, translated into Regulations in each Member state. In the UK, these Regulations are 'The Public Contracts Regulations 2015', or "The EU Regs" as we ('in the trade') call them.

The EU Regulations kick in when a contract exceeds a certain value but they do not have to be followed to the letter every time: there are circumstances when they do not wholly apply and our friends the buyers in private industry, for example, do not have to consider such regulations at all – the regulations that ensure such equanimity in one's approach towards purchasing only apply to tendering by public bodies such as central and Local Government, Registered Social Landlords, etc. (and for this reason, the transition from industry to public sector purchasing or *vice versa* can sometimes be harder than you would think).

The key thing to remember, though, is that the EU Treaty states quite categorically that, whether a procurement is exempt from the Regulations or not, a public body tender must be – and must be seen to be – non-discriminatory, transparent, proportionate and competitive. The processes and practices we shall cover in this book try to ensure just that.

One word of comfort. For some reason, people – especially those new to the game – fight shy of the '*bête noir*' that is EU procurement and fear it as they would some quasi-magical or other dark art. As we go on, you will find that, in reality, the only real differences between EU and any other procurement are the need to advertise in the OJEU[1],

1. OJEU – Official Journal of the European Union – see later

strict timescales for each stage of the process and some procedural matters that ought to cause you no problem at all. Just bear in mind that, if you do fall foul of the Regulations, the repercussions *can* be quite severe, but fear not – do it right and you'll be OK. Do not let the phrase 'EU tender' daunt you: treat it as 'business as usual'.

We shall look at the EU Regulations and their impact later on in this chapter. First, we shall consider procurement and procurement routes in general.

What is a Procurement Route?

A Procurement Route is the actual path or process you follow when carrying out a tender exercise to get a service provider on board – EU or not.

Tenders are conducted using one of a few set routes or processes that ensure the standards of fairness and competition are met and, because 'EU processes' are accepted as good procurement practice, the practical reality is that the same standard procurement methods are pretty much applied whether it is an 'EU procurement' or not. We shall therefore look at the standard procurement methods first and then afterwards see how the EU Regulations impact upon them.

In general, a procurement or tender essentially comprises five key stages:
i.   Advertising
ii.   Selecting bidders
iii. Tendering
iv.   Evaluating
v.   Awarding

Looking at these in a little more detail:

Advertising is how you let the market know that you are looking for firms or people to provide a service for you. Where you advertise and how (and even 'if') depends on various factors, and it is not always as straightforward as it sounds.

EU tenders will require an advert or Notice in the OJEU and non-

EU tenders will require advertising on your organisation's web site, trade magazines or the local press. If you are going to a number of bidders selected by another means (e.g. off an Approved List) – and you may for non-EU tenders – you will need to advise them of the opportunity and check if they are interested in bidding (this approach combines the advertising and selection process in one).

In addition to these options, the new Regulations require you to advertise in Contracts Finder every opportunity that is advertised (i.e. advertised anywhere).

Selecting bidders is where you consider firms' suitability, capability and capacity to carry out the work you are tendering and thereby their eligibility to have a bid considered. This process is carried out using a questionnaire specifically designed for the job and Chapter 5 looks at the matter of Questionnaires in some detail. You should certainly read Chapter 5 but, in the meantime, there are some general points you will need to know when looking at Procurement Routes.

The questionnaire is divided into two distinct parts:

i. The Corporate Questionnaire (so-called because it asks questions relating to your corporate standards) tests a firm's *suitability* to work for your organisation. It measures a firm's probity and commercial integrity and their standing against corporate criteria and legislative requirements. It asks for information on the liquidity of the business, its Health & Safety processes, its approach to Equal Opportunities, evidence of insurance cover, and so on.

ii. The 'Technical', 'Contract-specific' or 'Supplementary Questionnaire' (all the same thing – take your pick!) is used to assess a firm's *ability and capacity* to do the job itself. It asks questions about the company that are relevant to the contract being tendered, for example previous experience, level of resources, etc.

The two questionnaires (or two parts of one questionnaire) together enable you to determine which firms meet your corporate criteria *and* who have the ability to do the work.

Now is the time to see who can convince you they will do the work the best and at the most competitive price – the tender stage.

It is worth noting here that the Open procedure does not allow you to pre-select bidders as part of a separate stage ahead of the tender: you look at suitability at the same time as the tender, but we'll cover this later on.

Tendering is the stage where firms are asked to submit their tender or bid. A bid can comprise price alone – the cheapest wins it – or a combination of price and quality; that is, price and *how* they can do your job better than all the other bidders. This latter is called *MEAT (Most Economically Advantageous Tender)* and we shall see why later on.

To clarify a couple of key points about questionnaires and tendering:

i.   Questionnaires tend to look backwards and are about *the firm* wishing to bid – informative and generally non-competitive (except you may select the *most* suitable – see later).
ii.   The *tender* looks forwards and is about *your contract* and how they will do it and how much they will do it for – cost and quality.
iii. Remember – by law, you cannot ask at tender stage anything you have already asked at questionnaire stage (you will see why this is important later).
iv.   The Regulations (Regulation 59 to be precise) require you to accept a standard questionnaire known as the European Standard Procurement Document (ESPD), but enactment of this is still in its early stages. More of this later on.

Evaluating is the process by which you decide who is best to do the job, based on what they submit in their tender. Regardless of all the clever terms and conditions and specifications, it is the evaluation model – and the questions it asks – that will get you the best provider on board. It may look the smallest part of all the paperwork, but it is the one on which you should spend most time.

Once the tenders are opened, the process of evaluation begins. This can be easy or complicated, depending on what you have asked bidders to submit. If the bid is on price alone, it may be fairly rapid, and a QS may be able to carry out the evaluation in a matter of days. It may be delayed by errors or other ambiguities that the QS will need to clarify with the bidders and they will always need time to prepare a Tender Evaluation Report, but all in all this is one of the easiest forms of tender submission to assess.

If you have quality elements in your model, it will take longer as [a] the submissions are harder to score and [b] more people are involved. Once all the scoring is complete and moderated, the balance of price and quality must be worked out in the relevant proportions to arrive at the final outcome.

If the project is really complex, the evaluation of a price and quality bid may involve many people with different areas of expertise and collating and moderating the outcomes can take a long time. But remember – this is the stage at which you choose the company that is going to work on your project, so you need to get it right.

One key point about evaluation and evaluation models: always make it clear to bidders from the very outset how you are going to evaluate the bids (i.e. how you are going to score them) and always do it precisely in that manner. We said this about the Questionnaire stage – it is even more important at tender stage. Get it wrong and you could end up in court – literally!

Awarding is where the winning tenderer is actually 'given' the contract, although in most public organisations and certainly under EU Regulations, 'given' does not quite accurately reflect the hoops and hurdles the award process involves.

In principle, your organisation's Contract Standing Orders (or equivalent) will tell you who has to sign off a contract award and the process involved in getting there. You will recommend the winner of your tendering process in a Report of appropriate gravity (and length and complexity) and submit it for Approval. It may be a small project, and a half-page report and one signature may be enough. If it is an EU procurement it will be a large project and will certainly require Cabinet Approval (or your equivalent, e.g. Executive Board) and the report may have to have comments from up to six different specialist sections[2] advising on and supporting the recommendation before it can be submitted.

Such Reports can take a while to prepare: you will need to compose it, obtain contributions from interested parties and book it on the appropriate meeting agenda (there is often a two or three week lead-in period, maybe more). It can then be 'heard' and hopefully approved. Once heard and signed, there is normally a five-day call-in period before the Approval can be acted upon, and this can only *start* after the minutes of the relevant meeting are published – which in turn can take a week or more. I hope you are keeping up with this.

An EU procurement award process will also have to observe a mandatory 10-day standstill period but it is sound practice to observe one even if the procurement is not an EU process. We will look at this standstill period in more detail elsewhere – it is a sticky one – but if no-one objects to your decision (i.e. challenges you), you can proceed to contract (i.e. *actual* award) after all this is done.

It can all take a while, so *make sure* you allow the correct amount of time for all the constituent parts of the award process in your timetable. In addition, if it is going to be a Cabinet or Board Report (normally determined by its value), you may need to make sure it is registered on a Forward Plan, and this may need to happen up to three months before the meeting date – again your Constitution will determine this, but see how your timetable – doing things at the right time – is critical.

Then that's it – done. You can now go on to the next one...

Having understood all this, what follows next will be easier and make more sense.

---

2. Contributing sections may comprise Legal, Finance, Central Procurement, Risk, and so on, depending on your Organisation.

Procurement Routes

The 2015 Regulations identify five procurement procedures:

- Restricted Procedure
- Open Procedure
- Competitive Dialogue
- Competitive with Negotiation
- Innovation Partnerships
- The Light Touch Regime

The Restricted Procedure

As its name implies, this route allows you to restrict, through a pre-selection or qualification process, the number of people you invite to submit a tender. The pre-selection stage assesses the suitability and capability of a firm to bid by using a questionnaire comprising the two parts explained above. Because firms are selected in advance of the tender process the Questionnaire is called a *Pre-Qualification* Questionnaire (PQQ).

Once the contract is advertised, firms who think they might wish to tender request a PQQ, and it must be ready to be issued immediately. If they wish to *Express an Interest* in tendering, firms will complete the Questionnaire and return it by a given deadline. This period is therefore known as the Expression of Interest stage (although often referred to as the 'PQQ stage') and if firms request the PQQ late in the submission period they still have to observe the stated deadline. The returned Expressions of Interest (the completed PQQs) are assessed and firms that meet all the criteria are eligible to be *considered* for the final tender list – the list of those firms who will be invited to bid.

Note the use of the word 'eligible'. The Restricted procedure allows you to set overriding criteria limiting the number of people who will be invited to tender, regardless of the number that actually 'pass' the Questionnaire stage. You can stipulate, for example, that the highest-scoring eight firms will be invited to tender, or you can say that only firms with a score above 70% can go on to tender (even though companies with a score of, say, 60% might have demonstrated quite suitable credentials).

Through this restriction, you can limit the number of bids you have to evaluate at tender stage (and the number of firms each bidder will be competing with), the only key requirement being that you *must* make such conditions abundantly clear through the advertisement and in the issued documentation. If it is not clear, you will be in trouble.

We shall look at drawing up a PQQ in detail later on but you need to be aware that it has to be complete before it is issued. 'Complete' includes the scoring mechanism: your evaluation criteria must be decided before the PQQ is issued and must be declared within the document or made available through a process identified within it (e.g. web site address, for example).

The whole point of the PQQ process is to assess the suitability of a firm to work for the organisation and its capacity to deliver the service required. It is *not* a competitive process apart from your final selection criteria and it is *not* a tender. Firms need to provide factual information and are either suitable or they are not, and those who are not suitable will not be eligible for invitation to tender.

Be aware that the Regulations allow and encourage 'self-certification' by Applicants insofar as they can assure you at PQQ stage that they have the required insurance cover/accreditation/ etc. and you only need to ask for the actual documentation, by way of proof, if and when they get to the stage of award. Again aimed at smaller firms, this is designed to make Applications easier to submit but I am not sure that submitting such documentation is much of an additional chore and can see issues arising if you should get to the stage of possible award and find that the 'successful' bidder cannot meet the requirements after all. However, that is what the rules say so we have to allow it[3] and the wording of the document has to reflect this.

It is common for firms, once they see the PQQ, to have questions

3. Cabinet Office Guidance has been issued on this point.

either about it or about the contract it caters for. It is essential that answers to all such 'requests for clarification' are sent to everyone who has requested or who does request a PQQ. To achieve this, it is essential to keep a record of all those who have requested a PQQ and all the clarifications in such a way that responses can be easily distributed to all interested parties.

Your method for doing this will depend on the manner in which the PQQ is distributed and so you need to decide this. E-mail makes the process easier and electronic tendering systems solve the problem completely – but as I said, we shall come on to that later.

To facilitate the clarification process, the PQQ must contain details of how requests for clarification should be submitted and to whom; it should also set a final date for submission of any such requests and state the final date by which responses will be sent out. It is appropriate to not issue any clarifications less than eight days (or so) from the due PQQ return date so that everyone has a chance to amend their submission in the light of any new information. Responses can be sent out all together on the final date you specify or as they come in – that is up to you. You need to publish the query, ensuring that the enquirer is not identified, and enter the response alongside it.

On odd occasions, you may get a query that should not be issued to all (on the grounds of commercial confidentiality, for example, or a query regarding a firm being in the same group as another potential bidder). The PQQ should explain that all queries and responses will be distributed to all interested parties and that if a firm considers their question to be confidential in nature, they will have to state that at time of request. If you consider the response ought to go out to all, you will need to speak to the enquirer and resolve the matter. If you cannot resolve it, the query cannot be answered.

The basis of this is that all interested parties must have the same information and level of detail as everyone else. If any one firm has more information than any of the others, the core principle of equality of opportunity has been violated.

Remember that PQQs are NOT tenders – they are a submission of *facts* about the company. Nevertheless, they may contain commercially sensitive information and so need to be stored and processed in such a way that nothing they contain can get lost or interfered with.

Completed PQQs may start to drift in before the deadline because, for many larger firms, they are simple documents that they respond to on a regular basis and so they have the bulk of the information you require readily to hand. There is no reason why you cannot start to assess PQQs as they arrive, but if the project is such that a team of people are evaluating them or it is a complex process, you may wish to store them all until after the deadline and have a set period – a day or a week, maybe – for everyone to either come together or have access to the submissions and complete the evaluation process. Also, an Applicant may wish to amend their submission and, provided the deadline has not passed, this is fine. If you have already opened it and started work on it, this will mean you have wasted your time and could lead to questions around collusion (or providing unfair assistance) – it's best to wait.

You need to ensure you have allowed enough time for the evaluation processes to be completed and check that the corporate evaluation process does not need to be booked in ahead of time. Also, make sure you book the time in your evaluation team's diaries and that none of them are going to be on leave at that time. If there are problems with availability, make sure your procurement timetable allows for this. Always – plan ahead.

It is not unusual for applicant firms to make errors in their PQQ submission: they may send in or attach an out of date insurance certificate or Health & Safety Policy. The PQQ process is neither an exam nor a tender – it is an information-seeking exercise and it is quite in order to go back to the company, explain what they have done wrong and ask if they have the correct piece of information.

If you are short of interested firms, you will almost certainly do this; if you are inundated with able applicants, you may feel there is no need to 'nurse them along' like this. The golden rules of thumb are thus:

- You can pursue information or not – that is up to you, so long as you can justify your decision.
- This 'information' can be a specific document or clarification of a response.
- Whatever you do must be fair and equitable to all applicants – you cannot favour specific parties. What you do for one you must do, at least in equivalence, for all.
- Keep records of all communications relevant to any follow-up you make.
- Avoid giving more than one chance to anyone – persistent pursuit could be seen as biased assistance. Someone will always be watching.

Once the evaluation is complete, if a group of people has evaluated the same set of questions you should hold a meeting at which all the assessors agree on the final score for each question – a 'Moderation Meeting'. This ensures that no one marks a firm unfairly low simply because, for example, they have missed some vital piece of information. Through this, everyone is agreed on which firms qualify and go through to the next stage of the process – it becomes an *Evaluation Panel* decision. Do *not* simply average all the individual scores. On complex or high-value tenders, it is sometimes better to have an independent person chair this moderation meeting and it should always be someone who has not carried out any evaluation.

These days, most PQQs are issued and submitted (returned) electronically. If, for any reason (and there are some) the submissions are to be in hard-copy format[4], you need to be clear in the document where they are to be returned to, by when and how they are to be formatted and packaged. You also need to ensure

---

4. EU Regs require tendering authorities to be fully electronic by October 2018, Central Purchasing Bodies by April 2017

that the point of delivery you specify will be open and accessible at the appropriate time – check you have not inadvertently set a deadline for a Bank Holiday, for example (yes – it has been done!). It is normal to include a pre-printed label for their return.

Firms that do not pass the PQQ stage – or do pass but are not invited to tender due to other factors - should be advised accordingly and offered feed-back on where their submission failed to meet the required standard. Firms that do make it onto the tender list may be advised in the first instance but will anyway be sent an Invitation to Tender (ITT) in due course. Rarely will they want a post mortem on a successful submission.

Not all firms that request a PQQ will send in a completed Expression of Interest and by definition you will reject those who do submit a PQQ but do not meet your requirements. The Restricted Procedure is therefore the most commonly used procurement route because this initial vetting process limits the number of invitations to tender you issue and so avoids a tender evaluation nightmare by limiting the number of bids you receive.

This is the prime benefit, the drawback being that, having two distinct stages, it takes a slightly longer period of time to complete the process than its nearest 'rival' – the Open Procedure - and in some ways is more complex to manage. It is, however, better-suited to procurements where the assessment of capacity and technical capability might be a more critical or convoluted process or where you think the level of interest might be high and some initial 'weeding out' of interested firms will be of benefit.

The new Regulations also require you to issue 'all procurement documentation' with the PQQ. Unfortunately, they do not define 'all procurement documentation' so the jury is still out on that one – almost literally. This is an example where the courts will eventually arrive at a definitive meaning but in the meantime it is generally accepted that good drafts of all tender documentation should be issued with the PQQ. You should ensure that they are clearly marked as drafts and explain that minor changes may be made prior to issue at ITT stage and tat any such changes will be advised within the tender documentation.

This could no doubt incur delays in issuing the PQQ whilst the other documentation is being prepared for issue, but that is what has to be done. That's the Regs.

The Restricted Procedure is the standard or 'default' procedure but only for tenders *above* the EU threshold. Despite the advantages it holds for you the advertising authority it cannot, by law, be used for tenders below the EU threshold. When it is above EU, there needs to be a good reason for *not* following the Restricted Procedure – it really is the 'one size fits all' EU procurement route.

Because it is used above the EU threshold, some stringent requirements kick in and the Regulations lay down clear parameters for the questions you may use and refer to the EU Standard PQQ as the exemplar. This is something to bear in mind and we shall consider this when we look at drawing up questionnaires in more detail later on.

Enough advertising. Let's look at the alternatives.

The Open Procedure

The Open Procedure is a one-stage process and – as it says – is open to *anyone* who wishes to submit a tender. The Regulations require this process to be used for *all* tenders for contracts below the *Services* EU threshold (provided they are above £25,000 in value[5]) and this threshold applies whether it is a Services, Supplies *or Works* procurement. This latter point should be noted. Any works tender above the *Services* threshold can use the Restricted Procedure, even if it is below the EU Works threshold.

You can, of course, also use the Open procedure for any procurements above the EU Threshold if you wish.

By definition, there is no Pre-Qualification (or Expression of Interest) stage in the process and Questionnaires and tenders are issued and submitted as one package.

As there is no pre-qualification stage, the Questionnaire – although it asks the same questions – is called a Business or

---

5. £10,000 if you are tendering for a Central Government Department

Qualifying Questionnaire and is issued with and as part of the tender documentation.

In practice, there is no need for the Supplementary Questionnaire as all contract-related questions can be made part of the tender submission. Sometimes, however, it is felt that using a Supplementary Questionnaire is better than lumping everything into one tender document. For example, you can include some 'quick' pass/fail questions that will help weed out bidders who really do not suit your requirements and this is not best done in a tender document.

Whether you have a two-part Questionnaire or just the Corporate section, under the Open Procedure, *any* organisation that passes the Questionnaire 'stage' *has* to have their tender evaluated and that could be a lot of tenders. This is the key drawback of the Open Procedure – for the bidders and the tendering authority.

The requirement to consider the bids of all and any firms proven suitable rather than those firms proven to be *the most* suitable implies that all the questions in the Questionnaire must have a pass/fail scoring regime: an overall scoring mechanism would allow a low standard in one question to be offset elsewhere in the response. For this reason, all questions in the Questionnaire should be pass/fail and I would make it clear that a 'fail' on any one question will render the submission ineligible for further consideration. This will help reduce the work involved in assessing the Questionnaire returns. There is another way to reduce this burden:

It is normal practice to assess all the Business Questionnaires before the actual bids themselves are evaluated, as firms that fail to meet your Questionnaire requirements will not be eligible to have their tender evaluated. However, if the tender element has an 'easy' evaluation model, it is permissible to evaluate this bid part first and then look at the Questionnaire to check for eligibility and suitability to award. This is an option that may or may not make your evaluation process easier.

So you have some options. At the end of the day, it is your party and if you wish to set Technical Ability questions (provided they look *back*) in a part two Questionnaire, you may do so: just make sure the process you are following is made clear in the explanatory documentation that goes out with the tender.

Some firms prefer the Open Procedure as it is one 'simple' stage. On the other hand, some firms prefer the Restricted Procedure because they do not have to prepare a Questionnaire *and* a tender for their submission – they will only need to prepare the tender itself when they know they are eligible and suitable *and* they are one of only a relatively small number of bidders.

When using the Open Procedure, bidders will need to be given a little longer to complete and return the combined tender and questionnaire and also you cannot advertise the opportunity until you have both the documents complete and ready for issue, so any perceived time savings may not be as great as they at first appear. One obvious point to remember is, because all submissions are unavoidably part of a tender, you cannot open any (even to look at the questionnaire element) until after the submission deadline has passed.

The rule applies: only use the Open Procedure when the contract value is below the EU threshold (because you have to) or – if above the EU threshold - when you know specifically it is the most appropriate route to follow and the number of returns you will receive is low.

Competitive Dialogue

Specifically designed for larger and particularly complex procurements, this method is designed to bring on board industry expertise to work with the client during the tender process to arrive at a solution that meets MEAT criteria.

By its nature, the Competitive Dialogue process is flexible and with this comes an apparent flexibility in – or range of – terminology. To cater for this, I will give an overview of the process and then a brief glossary on the terminology that may be used at different stages.

Competitive Dialogue uses a Pre-Qualification process (as in the Restricted Procedure) through which a list of eligible bidders or prospective suppliers is determined. It is preferred (by the Regulations) to have three bidders left at the final tender submission stage, so bear this in mind when determining how many bidders to include on your tender list, but if the market will not allow this, then so be it.

Bidders that pass this stage are invited to enter straight into dialogue or submit an outline proposal based on the relatively scant information the client is able to provide at that time. I feel it is better to have an initial submitted outline proposal: bidders can be further eliminated at this stage if you wish and you have something on which to base the first stage of Dialogue.

A series of confidential 'Dialogues' are then held with each bidder, designed to further hone and refine bidder proposals so as to arrive at a solution that best meets the client's needs. Dialogue stages should be minuted and the notes agreed by all parties to avoid differences of opinion later on regarding what has been agreed and implied.

The number of Dialogue stages is determined by the client: it is normally two or three but it can be as many as it takes, and will probably depend on the complexity of the requirement; stages can be added if necessary. If the number of bidders is sufficient, it is usual to reduce at each stage the number that go through to the next by eliminating those whose proposals are not homing in on a solution – this saves both you *and* the bidder time and money.

Following each dialogue stage, bidders can be issued with an 'Invitation to Submit Detailed Solution', but I see that as loading extra work onto the bidders, when presentations at the start the next dialogue can serve the same process.

After the final stage of dialogue, a 'Call for Final Tender' (CFT) is issued to the remaining bidders.

Whatever you do, always make it clear in your initial documentation what the process will be and stick to that.

Finally, remember that even variations to the final submission can be tolerated once the contract is on site in order to achieve the desired result, provided the ogre of material change[9] is not stirred from its bed.

Competitive Dialogue (CD) is not a commonly-used process and the Competitive Procedure with Negotiation (see following) may see its use diminishing. However, CPN is not proving popular at the moment.

Competitive Dialogue is an expensive, complex and comparatively lengthy procurement route that should not be embarked upon lightly nor by anyone new to procurement. Fraught with legal pitfalls, Competitive Dialogue demands a lot of resources, needs a very good audit trail at all stages and is best left to more experienced procurement professionals. In all honesty – if you are reading this book you may not yet be ready to embark upon this process just yet. If you get the chance to assist or observe – seize it.

Mini-Glossary:
- Invitation to participate in dialogue (ITPD) – After bidder selection stage, invites bidders to participate in a dialogue process and setting out the terms applicable to that process. Does not imply the need for a submitted initial proposal at this stage.
- Invitation to submit outline solutions (ISOS) – Also after bidder selection stage, invites bidders to submit initial details of their proposed solution. The response may be evaluated, and this may result in a down-selection of bidders.
- Invitation to submit detailed solutions (ISDS) – This will typically be in the dialogue phase following an ISOS and invites bidders to submit full details of their proposed solution at that point in the process. The response may be evaluated, and this may result in the elimination of bidders.
- Invitation to submit final bids (ITSFB) – An invitation to the remaining bidders to submit their final bids at the end of the

dialogue process. (This is broadly equivalent to the ITT in a restricted procedure).
- Call for Final Tender (CFT) – Just the same as ITSFB
- Invitation to submit final tender (ITSFT) – Just the same as ITSFB

So what's the difference?

The difference in intent between the two procedures is not really explicit in the Regulations, but the Competitive Procedure with Negotiation is the more structured process. It requires a clear specification of deliverables – just as you would prepare for a standard, Restricted tender procedure – and a firm tender submission at the end of the negotiation process which becomes a contractual commitment. It is, after all, a competitive procedure *with* negotiation. The negotiation element provides leeway to, for example, value engineer the proposals, clarify costs in a rapidly moving market (e.g. excessive commodity inflation) and so on.

Competitive Dialogue, on the other hand, is a means to secure a solution to what is, essentially, a problem: you know what you need to end up with but do not know what is required to achieve it.

The 'specification' issued to those who pass the initial pre-selection stage can be as vague as it needs to be: even going so far as to purely indicate what you want to achieve at the end of the process, if that is all you know.

My personal view is that, by their individual nature, the Competitive Procedure with Negotiation (CPN) is exactly that – providing the opportunity to negotiate to a final tender on the basis of a requirement and an initial proposal. The implication of the term Negotiation supports the fact that detail can be discussed and 'final' offers adjusted on that basis (although this level of detail is not prescribed) but it would not necessarily be appropriate to use this method where major variations in the delivery mechanism (i.e. the *way* the outcome is delivered) need to be considered. It would be expected that the initial CPN offers would resemble the final offers and that major changes to the project would not be made.

By way of comparison, the Competitive Dialogue (CD) – as it says – is a dialogue, designed to arrive at a *whole solution*: the dialogue process will be based on a concept with required outcomes on which the dialogue and the final Evaluation Model will be based. The route to get from A to B can be virtually unmapped so long as what is delivered meets the need in terms of outcome, if not in appearance. In most normal circumstances, Competitive Procedure with Negotiation will be the way to go. If you think that CD is the best route then it may well be, but I would seek additional advice.

Negotiation without an OJEU Notice

Over and above the Competitive Procedure with Negotiation procedure, negotiation with bidders without going out to advert is allowed in *very* special circumstances and is strictly controlled by the Regulations. Negotiation can be exercised without advertising the opportunity when:

- no tenders/suitable tenders or requests to participate are received
- only one supplier could apply for artistic or technical or exclusive rights reasons (in the last two cases provided no reasonable alternative exists)
- extreme urgency from events **unforeseeable** by the authority, mean the time limits for competitive procedures cannot be complied with.
- products involved are manufactured purely for the purpose of research, experimentation, study or development

You cannot, of course, contrive to meet any of these qualifying criteria and, if negotiating under the first qualifying criterion, you must negotiate with *all* the bidders involved at the time the procurement process failed.

Innovation Partnerships

This is a very new process specifically aimed at Research & Development work and likely to appeal to smaller businesses, who often hold the cards in innovation and development. Whilst this procedure is obviously limited in its applications, it will be of great value in certain circumstances.

The idea is to encourage suppliers to work with and for the client to develop solutions not currently available on the market. The partners are procured using the Competitive Procedure with Negotiation (see above), and may be secured in stages to suit the process of development. Similarly, partners may 'drop out' of the development process as their required input reaches its conclusion.

On this particular procedure, the following should be noted:

- The rules of Competitive Procedure with Negotiation all apply – especially those pertaining to evaluation criteria, confidentiality and partner selection
- The ground rules must be very clear on Intellectual Property Rights and shares of benefits
- Ensure sound legal advice.

Light Touch Regime

I deal a little later on with the process of an EU procurement but include this here (even though it is specifically EU-related) as it is a procurement route that you need to know about and one which can be applied in specific circumstances. The Light Touch Regime is redolent of the good old Part B Services – a distinction done away with in the new, 2015 Regulations – whereby, even though the value exceeds the (specific) threshold, and the procurement is an EU process, the procedure has a very 'light touch' when compared with other EU processes.

The Services that can have the Light Touch Regime applied are specifically defined by their CPV Codes (see the Glossary) and relate primarily but not exclusively to social, health and other similar services. It is recognised that these services will have little, if any, cross-border interest even at higher values so they have a higher EU threshold than 'normal' services and, when their value is above that threshold, can be tendered using the Light Touch Procurement Procedure.

Here are 8 key points to remember about the Light Touch Regime and how to use it:

1. The Services to which the exception applies are now clearly defined through specified CPV Codes listed in Annex 14 to the 2014 Directive[6]. No grey areas remain.
2. These Codes apply mainly but not exclusively to areas of health care, education and social services.
3. There is a specific threshold (set by the EU at €750,000) below which the EU Regs do not apply to these Services at all; above this value, the EU rules apply but the Light Touch regime may be invoked.
4. The threshold, when translated into £ sterling, is currently £650,052
5. The Light Touch Regime requires the publishing of an OJEU Notice to advertise the opportunity and a Contract Award Notice (as per the old Part B) but...
6. The requirements for the conduct of the remainder of the process can be set by each individual member state.
7. The UK has adopted a very light touch for the bits in between – like the old 'Part B' requirements, which means you will have absolute choice of procedure (e.g. Open, Restricted), timescales and so on.
8. If the Service you are tendering is *not* covered by the CPV Codes defined in Annex 14, the 'normal' Services threshold will apply and the full EU Regulatory process must be applied to any procurement above that value.

---

6. A list and further guidance on these potentially exempt services can be found at the Tendering for Care website at http://www.tenderingforcare.com/uploads/document/doc/6607188/New_EU_Directive_and_Annex_XVI.pdf or in the Directives themselves.

That essentially covers the Light Touch Regime and will be heavily employed in certain areas of procurement.

Which one?

For most procurements, the straightforward Restricted and Open Procedures will serve yours and anyone else's needs.

The Regulations require the Open Procedure to be used in all procurements below the EU Services threshold[7], as it is believed this will open up the market to SMEs. This is a matter of conjecture, but Regs is Regs, as they say, and must be obeyed. However, when you have the choice, you should be careful when using the Open Procedure because it may encourage more bids than you can cope with: only use it where you have to or where you consider the response will be manageable – for whatever reasons. The time an Open procurement takes could be a little less than the Restricted Procedure but not sufficiently so to justify any additional problems that using the Open Procedure may bring about.

Competitive Dialogue and the Competitive with Negotiated Procedure are processes that follow an initial selection or pre-qualification stage (as in the Restricted Procedure) but allow for discussion with the bidders to finalise tender sums or agree how the service or project may be delivered to achieve the best results. Both are designed for projects that are more complex or risky, but Competitive Dialogue tends to be used for larger projects – such as building an airport. You will probably not be asked to do this. The Competitive with Negotiation will be useful in the guise of a two-stage Restricted Procedure for larger or more difficult construction or development projects and the like.

The Light Touch Regime replaces the old Part B Services procedure and is a much less onerous process for securing Services that meet the stipulated criteria. However, these criteria are strict and limit the opportunities when Light Touch may be used, and some of us may never experience its joys.

---

7. Note – the Services threshold. However, it is widely believed that this requirement will be amended in time up to a higher threshold.

Similar limitations apply to Innovation Partnerships where it will be a perfect fit for certain situations but not for general consideration.

The Restricted Procedure really is the 'one size fits all' process and ought to be your first or preferred option in most cases.

Procurement Routes – Useful Variations

The procurement routes described in detail above are the mainstays of tendering. We shall now look at some variations – or frills - that give you a degree of versatility and adaptability which can make your process perhaps more specifically fit for purpose. We shall look at them in no particular order and concentrate, for the moment, on tendering projects because [a] they provide the simplest example, [b] you will probably do more of these than anything else and [c] w shall look at services – including term maintenance contracts – specifically, later on.

The 2-Stage tender process

To illustrate this procedure, I shall give an example. A major, fairly complex construction project requires Planning Permission that may take many months to obtain, although the basic design is known and perhaps even approved to Outline Planning stage: it is the detail that is missing. Regardless, you need to get on and tender it.

In a two stage process, you would go out to tender with the designs and specification at the stage they are, and award preferred bidder status to one of the tenderers; that takes the relationship to the second stage where, with the designs complete and the specification finalised, the contractor firms up the price and submits their second stage tender. Some discussions may take place at this stage, akin to a dialogue but *not* a Dialogue, and the final price may or may not be accepted by the client. If it is not accepted, the final tender process reverts to the company who was next in line at the first stage.

The Competitive Procedure with Negotiation lends itself handsomely to this approach, but may not necessarily be a direct replacement: the difference is subtle, but it is there (primarily, the new procedure requires tenders from all bidders who participated in the process) and it may now be a preferred option.

The 2-Stage process is most beneficial and effective when used in conjunction with Early Contractor Involvement.

Early Contractor Involvement (ECI)
ECI is used where the contractor's knowledge will benefit a design process: the contractor is brought in before the design is finally complete and works with the design team on the final stages to advise on buildability, cost savings, materials, and so on.

Early market involvement – strongly advocated in the 2015 Regulations for any procurement – will allow you to consult with the market and float your ideas and this will certainly help you formulate your project and the route to market. However, you may rightly wish to get a constructor more directly involved at an early stage and this can be achieved simply by tendering the works earlier, as I have described above, and usually the 2-stage process is used for this. Sometimes you can tender the project on a Schedule of Rates that can be applied once the final designs are agreed; the tender method should suit the situation but the intention is to get building expertise on board to help with the design.

Sometimes it is even better to have the constructor on board from the beginning and to achieve this you have two choices: you can either tender for the designer and the constructor both at once or you can tender one Design and Build Contract. In the event of tendering the design package separately, you are referred to Chapter 16 and the section on Novation. Meanwhile, let us look at the Design and Build option.

Design and Build
You can tender for a company to provide you with the complete package – to design the project and then to build it and off-the-shelf contract suites provide for this process. Sometimes a construction

company will have a design team within it and sometimes they will team up with an architect's practice to provide the service.

The tender process is a combination of tendering for an architectural service and a construction project and there are issues around tendering for a design service. We shall look at these when we look specifically at tendering for services: for now, we are looking at project tendering options.

Lots and Packages

If your contract has 'parts' – perhaps it is a maintenance contract over several discreet geographical areas or it is to build extensions to three different schools - you can tender one contract with a separate 'Lot' or 'Package' for each distinct part. Using the school extensions as an example, each school project would be a 'Lot' or 'Package' within the one, tendered contract. Firms would be invited to bid, normally, for one, some or all of the packages and you would expect a company bidding for all three to give you a better price for each than a company bidding for just one, simply through economies of scale if nothing else.

The Regulations require you to package a contract into Lots where this is possible and support the principle of mixed availability of Lots as well as the reserving of Lots specifically for SMEs, so the option has to be considered, even for below-threshold tenders. In fact, the Regulations go further than this and say that, if a large contract is *not* divided into Lots, you need to explain why in your OJEU Notice. Give Lots some thought.

Your evaluation model must enable you to analyse which combination of what Lot/s with which firm gives you, the client, the best deal and you may wish to consider limiting bids in such a way that one Lot – maybe the smallest – is reserved for isolated bids in order to give a chance to smaller firms.

Be careful to ensure that the tendering 'rules' in your evaluation model are clear, and try to avoid over-complication.

More details of all these options will appear as different aspects of tendering are examined in more detail, but knowing the above will assist you, at tender planning stage, to identify the procurement route that suits your needs best and to timetable accordingly.

Tendering below the EU Threshold

I have described in general terms the processes involved in the various types of procurement routes and will go on to look at the impact of having to adhere to the full EU requirements in a moment. First, though, I will cover the requirements of the Regulations for tenders that are below the EU threshold.

The EU Directive did not cover this, but the Regulations do, because the Government included some recommendations made by the Lord Young Report aimed at improving the access to tendering opportunities by SMEs. I touch on these requirements throughout this book but summarise them here for added clarity.

1. These requirements apply where the value of the contract exceeds £25,000[8].
2. Only the Open Procedure may be used where the value of the tender is below the EU Services threshold (even if it is a works tender): there can be no pre-selection, only a questionnaire issued as part of a single-stage tender process.
3. The Questionnaire can assess the corporate requirements (financial standing, insurances, etc.) and suitability/capability (previous experience) but must comprise pass/fail criteria only.
4. You must be accommodating when assessing the financial information and allow SMEs and new companies some leeway when deciding if they present too high-a financial risk.
5. The limit of twice the contract value for a firm's annual turnover applies (see the section on financial assessments for guidance on this and point (4) above)
6. Bidders can self-certify on insurance and other accreditations and you ask to see the proof and certification only when at the point of decision to award

---

8. £10,000 for Central Government Departments

7. You can assess the tender submissions first and only consider the questionnaire element of the bidder/s selected for potential award.
8. The requirement to advertise in Contracts Finder *if* you advertise *anywhere* else applies, as does the requirement to post advice on the tender outcome.

Now we can look at procurements that hit the EU Threshold.

When is it EU?

The thrust of the EU Regulations is based upon the EU Treaty (TFEU) principles of non-discrimination, equal treatment, transparency, proportionality and mutual recognition. These aims of openness and fairness are core to any good procurement – EU or not – and set a benchmark of good practice. Thus, the *Treaty* applies to *all* tenders, EU or not.

The EU Regulations, on the other hand, are interpreted (in England, Wales and Northern Ireland) through the Public Contract Regulations 2015[9] and have to be observed by all public sector bodies (such as Local Authorities – really, anyone sending public money) when tendering a contract above a certain, specified minimum value. You therefore need to be sure about whether your organisation is classed as a public sector body or not and whether therefore you are obliged to follow the Regulations. Nevertheless, it is as well to note that many of the requirements laid down in the Regulations are generally accepted as good practice and bear consideration in any procurement.

The EU Regulations come into effect when the value of a proposed public-sector contract will *or even might* exceed certain specified financial thresholds. The thresholds are set in January every two years (2016 was the last occasion) and are based on the Euro currency, leading to odd values in £ Sterling. The (EU) Regulations divide contracts into certain, specific categories but the ones that will concern us are:

---

9. Scotland has its own Public Contracts (Scotland) Regulations

❖ **Works** – such as construction projects and major refurbishments

❖ **Services** – such as term maintenance provisions or consultancy work

❖ **Supplies** – such as goods or products

At the time of writing[10], the Works threshold is £4,104,394 and so any contract exceeding – or likely to exceed – that value has to be procured in strict compliance with the EU Regulations. For Supplies and Services, this value is currently £164,176 (with the additional threshold of £650,052 for Services falling under the Light Touch Regime, remember). You must be aware of the *current* thresholds to ensure you stay within the law.

The EU Regulations stipulate what and when contracts *must* be tendered under their regime but they do not forbid any contract that falls outside of this scope being tendered using an EU process if it is thought there is a need or a benefit in doing so. In most cases there will not be, but at least now you know.

What is the impact of tendering under the EU Regulations?

The answer to this, in practical terms, is "very little", so do not be put off by the fact that a tender may have to be undertaken using a full-blown EU procedure. Additional care may be required because the impact of bad practice can be quite major, but it is not more burdensome. Just do it properly.

There is a range of requirements laid down in the Public Contract Regulations 2015 (referred to hereafter as the EU Regulations or EU Regs) that might be considered additional to non-EU procurement processes. These key differences are:

1. You may use the Restricted Procedure
2. A PIN Notice pre-advertising the opportunity may be published

---

10. The threshold values are revised in January, every two years: these figures were set in January 2016.

3. You must publish an OJEU notice advertising the tender
4. Minimum timescales are stipulated
5. You must hold a Standstill Period
6. You must publish the details of the winning bid in the OJEU
7. You must maintain a Procurement Report
8. There are some other requirements within the documentation.

1.   *You may use the Restricted Procedure*

If the procurement is below the Services EU Threshold, you have to use the Open Procedure to provide better opportunities for smaller providers. Once you start tendering opportunities above this threshold you have the whole gamut of processes to choose from, including the trusted Restricted Procedure.

2.   *The Use of PINs*

A PIN (Prior Information Notice) can be issued at any time to let the market know that you are thinking of going out to tender. Issuing a PIN is *not* a commitment to procure: you do not have to tender that contract, so issuing advance notices like this still allows you the flexibility to change your mind or adapt your strategy but it does fore-warn the market, allowing you to assess interest and embark on pre-tender market engagement.

You will see later that publishing a PIN does sometimes allow you to reduce the tendering timescales, provided the tender you issue is the same as the opportunity described in the PIN: if you do tender but change *what* you are tendering from that advised in the PIN, the PIN time concessions no longer apply and the time-reduction benefits disappear as well.

Some organisations publish a PIN listing all of their intended procurements for the year ahead – this means the benefit of reduced timescales (see later) is always available (although why you would need them if you know about the procurement that far in advance I do not know).

One further use of the PIN is having it serve as a Contract Notice or 'Call for Competition'. You can use a PIN to pre-warn the market *and* serve as the OJEU Notice(see below) by declaring this in the PIN, but it does mean that, instead of a relatively high-level descriptions of the contract as you might do in a purely advance-notice PIN, you will have to give all the detail you would normally have given in the OJEU Notice. Also, PINs can serve as a Call for Competition for more than one procurement.

As firms respond and declare an interest, you have to keep a note and contact them to advise them when you are about to go to market. The benefit is earlier market notification and a longer time for the market to consider the opportunity; the un-benefit is having to make sure you keep a track of any expressions of interest (especially if you are advertising more than one opportunity in one Notice).

One more thing to remember – you cannot use a PIN as a Contract Notice *and* make use of the time reductions shown above – it either has to serve one purpose or the other, not both.

3.   *Publishing an OJEU Notice*

If you use the PIN as a call for competition, the time when you contact interested parties to advise them the tender documents are ready for downloading marks the start of the procurement process proper. If you do *not* use the PIN as a call for competition, the first (official) stage in any EU tender is the placing of an advertisement, Called a Contract Notice or 'OJEU advert[11]', in the Official Journal of the European Union (OJEU).

The OJEU advert is submitted using a standard, official pro-forma, normally (and advisedly) accessed through one of several official websites. There are various organisations (or 'portals') that will issue an OJEU for you (normally for a fee), or you can do it yourself via the official EU website SIMAP[12] and most e-tendering systems enable you to issue OJEU Notices direct.

OJEU notice pro-formas can be accessed on SIMAP through the link to "e-notices" and you can complete the form, or start it, save

---

11. We will look at OJEU notices in another chapter.
12. SIMAP web address: http://simap.europa.eu/index_en.html

it on the site and return to it later to finish and then issue it when you are ready.

The OJEU Notice is, at first sight, a daunting document and can take a while to complete. I will not use space here giving the detail of how to complete one: if you have not completed an OJEU Notice before you will almost certainly need some guidance and I would recommend you ask your legal support to give it the once-over. On the SIMAP site is access to another facility called TED – Tenders Electronic Daily. You can enter this facility and see OJEU Notices issued by other bodies similar to your own and get an idea of what information is wanted and where and how it is entered on the form. You may even search and find one of your own organisation's notices from a previous tender exercise and you could crib a lot of the standard information from that.

However, here are some important points about OJEU Notices:

- The OJEU advert or Contract Notice requires a description of the contract and how it is to be tendered, as well as specific items of information regarding the tendering organisation, etc.
- Some of the things you need to fill in are obvious, such as the procurement route you are using, submission deadlines, and so on.
- Some of it is standard to your organisation and repeated in every Notice – use previous Notices as a guide
- Whatever you put in the OJEU is a legal commitment (for ever) and you cannot digress from it in any way. If you specify the contract is for street lighting, you cannot use the appointed provider to carry out any white lining, for example. You *must* be clear in your mind what you require – and *might* require – from the contract and the provider in the future so that you do not hog-tie yourself with a contract that does not fully meet your needs or give you the scope you may need in the future.
- If you *do* allow the contract to deviate from what is detailed in the OJEU (scope creep can be a risk with contracts) this

could become what is known as a 'material change': you are commissioning what has not been legally tendered on the open market and in fact can leave yourself open to legal challenge. Beware – Material Change presents a real risk.

- The contract description is supported by numbers called Common Procurement Vocabulary Codes (CPVs). These are again available via SIMAP and *very specifically* define the needs of the contract. You must make sure you get these codes right and include in your Notice all of the ones you are going to need. Firms (or their agents) search for tendering opportunities using these CPV Codes and if you leave any out you may miss a good provider off your tender list. At the same time, these CPVs define the scope of the contract, so I refer you back to the second bullet point above – make sure you cover all the categories you may (or may possibly) require over the life of the contract.
- You must include details of any Lots or Packages[13] and the projected value of the contract. You do not need to be precise and some variation at time of actual tender is allowed, but you need to make it accurate enough for potential bidders to know if they are interested and capable.
- You need to indicate how you will be evaluating tenders and PQQs either in full or as an overview, referring to the tender documents themselves for the detail (recommended).
- To avoid excessive detail or volume of text in the OJEU, it is normal to refer interested parties to information contained in the PQQ or tender documentation itself. You can also issue (if the complexity or scale of the project demands it) a Memorandum of Information giving a more detailed explanation of the opportunity and what you will be looking for from bidders.
- The Contract Notice must indicate any restrictions on tendering – e.g. for mutual bodies or sheltered workshops (see Chapter 13 on Sustainability).
- The Contract Notice must indicate when you expect the contract to be awarded or start and how long it is intended to last.

---

13. See Chapter 6

- Always have someone knowledgeable sanity-check your OJEU Notice before you send it off.
- You can correct errors in a published notice or change the information by submitting an Amendment – which is really just like another OJEU. However, once you have completed an OJEU Notice, you will realise why you really want to get it right the first time. If you amend a Notice, the timescales clock starts again from zero.
- You can cancel a tender process after it is advertised but only by placing a Cancellation Notice in the OJEU.
- The Expression of Interest (EOI) stage starts on the day the Notice appears in the Journal.
- The Notice can take a few days to actually get published, so add a margin onto your timetable for this – say three days. OJEU will advise you when it has gone out.
- You can extend the EOI period by entering a longer period in the OJEU Notice or (if the need arises) after it has been published by issuing an Amendment (without going back to time zero).
- Any changes after the OJEU Notice is published – even if you have published an Amendment Notice - must be communicated personally to all those who have already expressed an interest as they cannot be expected to identify and read the new notice.
- You will be able to read your own published advert on TED.

This OJEU advert is the key difference between an EU and a non-EU tender. As you get more practiced, and the OJEU gets less daunting, your fear of EU procurement will get less to the point of zero.

In addition to the OJEU advert, the Regulations require a matching advert to be placed on the Government's website *Contracts Finder*[14] to ensure visibility to SMEs. This is laudable and an e-tendering portal linked to Contracts Finder will make this process easier and less of a duplicated effort.

It is worth noting here that an advert has to appear on Contracts

14. https://www.gov.uk/contracts-finder

Finder *whenever* you advertise an opportunity that has a value of £25,000[15] or more – whether the advert is an OJEU advert or just an ad in the local paper. If you do not advertise the opportunity (e.g. by selecting from an Approved List), then you do not need to place a notice on Contracts Finder.

You should always advertise the opportunity publicly in suitable media. The obvious and usual places to advertise are your organisation's website, the local press (to provide potential opportunities for local businesses) and a recognised trade journal or other national publication. The public (media) advertisement, whilst obviously in a different format, must reflect exactly the details in the OJEU advertisement and make clear how to obtain a PQQ and by when it has to be returned. Sometimes advertising on an organisation's website is considered adequate without any of the other options, but I am not convinced. Essentially, you need to reach out to the market that can provide what you want so you need to know what you are seeking and direct your publicity accordingly.

Whichever medium you use to advertise the tender, any public advert **must not** be published (i.e. appear) before the OJEU Notice is published. This is a legal requirement. You therefore need to research the copy and publication dates of your chosen journals and ensure the OJEU appears first, but not by too big-a margin. If your trade journal appears too long after the OJEU publishes it may not give those applicants a reasonable time in which to complete and return their PQQ. In cases where this happens, in the interests of fairness and equality, you will need to either delay publishing the OJEU Notice (and so the return deadline) or have an extended PQQ stage to allow all participating firms adequate time to submit.

You will need to determine your press dates early (as part of your tendering strategy) and include this advertising stage in your timetable otherwise you may end up with a problem before the tender process has even started.

---

15. This value applies to Local Authorities and their equivalent. For Central Government, the value is £10,000.

If the tender is non-EU, a Notice does not need to go in the OJEU but you do need to carefully consider your advertising media to achieve maximum effect. Sometimes, as I have already alluded to, it is considered beneficial to advertise your opportunity in OJEU even if the Regulations do not strictly require it. And don't forget Contracts Finder!

Whatever medium you use, the appearance of the first advert marks the beginning of the Expression of Interest stage.

4.   *Timescales*

With non-EU procurements you have freedom to determine how long to allow for each stage in a tender process; when it is EU, the Regulations are quite specific about the minimum time that must be allocated to each part.

The stipulated minimum timescales are as follows:

Procedure	PQQ – calendar days			Tender – calendar days		
	Normal	Via e-tendering portal	With PIN	Normal	Via e-tendering portal	With PIN
OPEN		N/a		35	30	15
RESTRICTED	30	30	15	30	25	10
COMPETITIVE WITH NEGOTIATED	30	30	15	30	25	10
INNOVATION PARTNERSHIP	30	30	15	30	25	10
COMPETITIVE DIALOGUE	30	N/a		N/a		
ACCELERATED OPEN		N/a			15	
ACCELERATED RESTRICTED	15	N/a		10	N/a	

Figure 3.1 – EU Tendering Timescales

It will be seen that:

- There is no PQQ stage for the open Procedure – but you already know this.
- The minimum tender-period timescales can be reduced for some procedures if all of the tender documents are readily

available and submissions can be made via the internet (i.e. using an e-tendering portal).

- Both PQQ and tendering timescales can be reduced by the publication of a PIN Notice. However, you cannot take advantage of this concession if you tender the contract less than 35 days after it has been published or more than one year after. This 35 days' wait means you may not gain anything in time, but a PIN can serve to alert the market to your intentions to tender, assess market interest and serve as a prompt for industry consultation or pre-tender market engagement.

- You can adopt an Accelerated Procedure if it is a genuine emergency. A genuine emergency can be defined as securing measures to deal with the effect of fire or flood, for example. Starting the procurement too late is not considered suitable justification for using this procedure.

- The regulations allow you, when using the Restricted and the Competitive with Dialogue Procedures, to reduce the tender stage period down to a minimum of 10 days provided all bidders agree to the proposal. If you suggest 10 days and one bidder asks for 15, then you will have to settle on 15 – but again with everyone's agreement.

- Overall, despite people's concerns, following a full, EU-compliant process does not necessarily delay the tendering process. What causes delays is nearly always issues such as the specification not being ready or a delay at an evaluation stage.

5.   *Standstill Period*

The Standstill Period is covered in great detail later on. Suffice it to say here that you have to allow a 10-day period (minimum) after the evaluation is complete and before the contract award is made when the bidders, having been advised of the outcome, have the opportunity to seek further information on their submission and get briefed on why they were not successful.

It is always advisable to recognise a voluntary Standstill Period for non-EU tenders, but it is compulsory for above-threshold procedures.

6.   *Contract Award Notice (CAN)*
Lastly, you must publish the outcome of the tender in OJEU using a Contract Award Notice. The CAN must be issued within 30 days of the award being made or you can include it in a quarterly 'return' of all awards made in that period.

You must also publish the outcome in Contracts Finder (although this applies to advertised contracts below the EU threshold as well).

7.   *Maintaining a Procurement Report*
You should maintain a record of the procurement process. This is not a diary but a record of the process that will serve to explain events and key decisions (or reasons) along the route. This must be kept to hand and may be called upon by the Cabinet Office or the Courts if there is reason to suspect bad practice or if an explanation for a particular course of action is required. These reports must be retained for three years from the award of contract.

8.   *Other requirements*
The following are some of the other points that need to be borne in mind when carrying out an above-threshold tender. They are important, of course, but they do not impact on the process, as such, more on the matter of documentation.

a) *Exclusions and Eligibility to Tender*: There are situations where companies are not allowed to tender and your *have* to exclude them from the process; there are circumstances where companies *may* be excluded from the process and the choice is up to you. Regulation 57 lays down the criteria against which these decisions are made and they centre around convictions of fraud, tax evasion, use of child labour and all manner of misdemeanors. The list within the Regulations is long and I shall not reproduce it here but the recommendation is to familiarise yourself with the list. In practice, the Standard

Questionnaire asks bidders to confirm that, under Regulation 57, they are eligible to tender. It is often useful to actually issue a copy of the requirements of Regulation 57 in full for information purposes and this can be attached as a standard EU tender document covered by a specific clause in the Questionnaire.

b) *Contract Termination Clauses*: You are required to put into contracts clauses that allow you to terminate the arrangement in the event that you are required to do so on grounds that lie within the Regulations. These grounds can include the contract being declared ineffective (see below) or it being discovered that the provider has been convicted of one of the misdemeanors described above. Normally, such clauses will already be contained within your organisation's bespoke contracts or standard amendments.

c) *Grounds for Ineffectiveness*: If the procedure is an above-threshold procurement, it is covered by the rules on ineffectiveness (see Chapter 11). Tenders below the threshold are not covered by these Regulations.

d) *Reserved Contracts*: Certain types of contract can be reserved for tender by Social Enterprises, other not-for-profit organisations and sheltered workshops. The categories eligible for reservation are listed in the Article 77 and there is more of all this in Chapter 13 on sustainability.

e) *Standard PQQ*: There is a standard PQQ issued by the Cabinet Officer for general, mandatory use. You are required to use the style and/or format of its questions wherever and as appropriate and always for above-threshold tenders, adapted for the Open Procedure as necessary.

f) *Availability of Documents*: Whatever procurement route you choose, you must issue all available documentation and any documents referred to within the OJEU advertisement, unless you have a good reason not to (e.g. documents have to be issued via specific, incompatible software or if a model is required). If, because of one of these reasons, you are unable to meet this requirement you must add 5 days to the minimum EU timescales. The key thing to remember is the phrase 'all available documents'. If they are not written, they do not have to be prepared specifically to be issued with the OJEU Notice. If they are prepared, or part-prepared, you should issue them with the nomenclature '*DRAFT*' clearly added.

This section posed the question 'What is the impact of tendering under EU Regulations?' I have noted 13 key areas where there are things to do that you might not do in a non-EU tender. It sounds a lot but, if you look at each one on its own, none of them are onerous. Carrying out tenders above the EU threshold is not foreboding – or shouldn't be. Take it easy, plan it properly and all will be well.

Summary

- The two main procurement routes are the Restricted and the Open but they are not the only – nor necessarily the most appropriate – ones.
- EU Regulations mainly govern timescales for specific stages in the procurement process.
- EU timescales are minimum; they can be extended but cannot be reduced other than by processes laid down in the Regulations themselves.
- There are allowable variations to the Procurement Routes that enable you to better-tailor the process to meet your specific requirements.
- The new Regulations are more complex regarding sub-EU threshold tenders and the use of a Questionnaire, so tread carefully in these areas and watch for further Cabinet Officer Guidance.
- Failure to comply with the EU Regulations is a very serious misdemeanour – don't do it!
- EU procurements generally all require a Contract Notice and Contract Award Notice in OJEU – but there are exceptions.

- Don't be intimidated by the prospect of carrying out an EU procurement!

4 // Planning the Procurement

The Need for a Procurement Plan

All projects need a plan. If the plan is bad – or if there is no plan at all – the project at best will not have a good outcome and normally it will just fail miserably. Procurement exercises are all projects, albeit complex projects, and should be treated as such. They have an aim, a beginning, a middle and an end (which ought to be the same as the aim) and in most cases the outcome is critical to an organisation's service delivery, its reputation and its budget so, yes, it is pretty important to get it right.

A procurement plan (or strategy) has to be as detailed as you need it to be and take on board all the issues that relate to the tendering process and beyond. If you do not have to present this plan to an audience, it can be in any form you like: notes, bullet points, tables – you name it – but do have it in a format that you are comfortable with, that you can keep to hand and can modify as necessary to meet changing circumstances.

It is important that everyone involved in the procurement (the 'Team'[1]) knows the plan and approves it. For this reason, they all have to be able to understand it, so bear this in mind when deciding on a format and involve everyone in drawing it up – make it the Team's plan. 'Ownership' springs to mind.

If you have to submit a pre-procurement or strategy plan to a Senior Manager or Director for their Approval, then it will probably have to be a report in a specific format that will include the main thrust of your plan, but not all the minor (albeit vital) details that you will need to remember.

If you are working to PRINCE 2 principles, then the PID[2] will be a very detailed Pre-Procurement Plan that will leave nothing to the

---

1. We will look at Project Teams later in the book.
2. PID – Project Initiation Document – a very detailed Project Plan that requires approval and sign-off before the project can proceed.

imagination. Normal project plans – or your procurement plan - will fall somewhere in between the PID and a scrap of notes.

I feel I may need to distinguish here between a plan and a report. A Strategy Report is for others to read and approve and will summarise your plan, probably emphasising the business case and demonstrating the viability of your proposed strategy. In general the report will have to be (and needs only to be) as informative as your Approver wants it to be; it is common in Local Authorities for a Report to have comments or concurrent paragraphs from other key departments (legal, finance, etc.) to confirm that their particular aspects of the matter in hand have been considered and any issues covered. In this case, it is your procurement strategy; later it will be the recommendation for award.

How you impart this information (i.e. the format of the report) will depend on your particular circumstances and the environment in which you operate – I will leave that bit up to you – but if you need to submit a strategy (or other) Report, you will need to source the template and find out who has to add 'bits' to support your proposals.

It is timely to point out that reports may have different formats for different levels of approval. Your organisation's standing orders will stipulate who has to approve your proposal (normally this is dependent upon value amongst other things) and you should find this out before you begin. You *cannot* tender anything – or commit your organisation to any level of spend – without the appropriate authority to do so: ignoring this is generally a very good way of becoming unemployed. So make sure that a copy of your Standing Orders is to hand alongside a copy of this book and read both – avidly.

You also need to consider internal *processes*. Some organisations have a Gateway or Tollgate process that seeks reports at certain key stages and approves them – or not – before they go 'up' for formal ratification. If your organisation has such a process, adequate time needs to be allowed within your strategy as passing a Gateway stage can sometimes be troublesome. Often, Reports can only be received on specified dates so again, your timetable needs to allow for this.

The (real) procurement *plan*, as opposed to the Report, is yours, however, and should include all of the things *you* need to consider at the start of and throughout the process. It should include – as a minimum – the following:

Reason or Business case	The purpose of the exercise – what you are trying to achieve and why.
Overview	How this project may effect improvements and address issues (almost a reminder)
The chosen procurement route	Why are you doing it this way?
Timetable from 'now' to 'start on site'	To ensure the contract will start on time (the most important part)
Form of contract and period	May not yet have been decided so there may be options.
Budget	Funding arrangements, cost issues, etc.
The Project Team	Who will be working with you on this procurement including who needs to be commissioned from outside the organisation.
Consultation	With whom, why and roughly when
Risk Register	Primarily for the procurement process but also includes contract operation risks that the procurement process hopes to mitigate (see Chapter 12).

Figure 4.1 Procurement Plan Content

We shall look at each of these parts in turn but if, by the end of it, you still feel unsure of any of the issues raised, don't worry - all of them will become clear as we continue to cover the procurement process.

Looking at each point in turn:

The Business Case

"*Why* are you carrying out this procurement?" This is a very simple but fundamental question and one that needs asking because it is the driver for the whole process. I have already referred to the importance of the 'why' factor in procurement (and will continue to do so) and in the very first instance there has to be a reason or 'business case' that identifies who has the need, why the organisation needs to meet it and an outline of how it is intended this need shall be met. These three points are at the very core of your procurement process. They will, in reality, define the intended *outcome*.

How the contract will be paid for is also relevant, although in most cases, confirming that a budget has been identified (Capital, Revenue, Grant, etc.) should be sufficient at this early stage.

Just bear in mind that the business case itself is more important for Reports (see above) than your own Plan. All *you* need to know is the 'who' and the 'why' so that you know what your tendering exercise needs to achieve.

Often the reason for a tendering exercise is quite straightforward. In Housing, for example, there are many ongoing statutory duties that a Social Landlord is obliged to carry out and these duties are increasingly being met through outsourcing. Increasing externalisation means that all or most of these services have to be tendered on the open market in adequate time for a new provider to take over at *precisely* the moment the old contract expires. Case proven.

Overview

Written simply as a list of broad objectives and things to remember, this is an opportunity to make a note of any points relevant to the procurement that will need to be taken on board as the process develops. It may include where you want (or are wanted) to make service improvements, any budget limitations, known changes to asset lists, etc. At this stage, the points noted here will help feed into the rest of the plan and the procurement process itself.

The Market

The state of the market that will provide the service impacts upon your procurement strategy or plan. If the market is buoyant it may mean that prices will be higher; it may also mean potential providers are busy and less keen to tender and you may have to temper your procurement strategy to suit.

Oddly, it can be the case that a rising or buoyant market can induce firms in that sector to overstretch themselves or their resource costs to increase such that they become vulnerable, presenting a higher risk than you might actually suppose. Assessing the market enables you to consider these factors.

The chosen procurement route

There are a variety of procurement routes, and you will need to select the one best suited for your purpose. In addition, it may be EU or non-EU. A procurement becomes 'EU' when its value exceeds a specific sum laid down in law by the European Union and specific rules and regulations apply to the tendering process. However, it is nothing to worry about and has been dealt with in Chapter Three.

Once you are familiar with all the options, you will be in a position to identify and justify the chosen route and provide for its processes in your procurement timetable.

Timetable

Drawing up the timetable is often found to be the stickiest part of the planning process and is even considered by some to be the hardest bit of the whole procurement. Unfortunately, despite its difficulty, preparing the timetable *is perhaps the most important part* of the whole tendering process. Remember that.

There are two key points to remember about a procurement timetable:

- You *must* start the procurement process early enough to complete the exercise properly.
- You *must* allow sufficient time within the process for all stages.

Starting on time

Many people are often keen to display their procurement prowess by claiming they need alarmingly short periods of time to carry out a tendering process. Don't be fooled or feel insecure or inadequate. On paper, amazing things can be achieved but in reality there is an abundance of activities that will take time to complete properly. I can list a few, all or any of which might apply to your contract: writing the specification, consultation, TUPE, internal approval processes – the list goes on. Appendix 1 gives a more specific list of some of these activities and the periods of time they may take. If you start the tender too late, [a] it may be impossible to complete on time and [b] it will certainly be too late to ensure the quality of the product you create.

There are ways to avoid this problem. One is a Forward Procurement Plan that will advise what contracts are coming up and when the tendering process ought to start (and I include the whole procurement process within the definition of 'tendering'). Another way is to ensure that whoever administers or manages a contract knows when it is due for renewal and lets you know so that you can start the next procurement in good time. Lastly, if it is not a contract *renewal*, do not make promises about the procurement period that you cannot possibly keep.

A procurement can be expedited through various means, and we look at these elsewhere, but these can only be used in specific circumstances and are not designed to compensate for poor planning. If people push you for a quick result, make this clear.

Drawing up your Timetable

Timetabling gets easier with experience but you need to be suitably 'equipped' to draw up a meaningful timetable and be aware of all the stages that will make up the whole procurement process. Appendix 2 shows a typical (for me) timetable that identifies all the key milestones in a procurement within a specific administrative environment. *Your* administrative environment may be different, but do not worry about that. Do not worry, either, about what each line item actually means. By the end of this book you will know *exactly* what each one means. This diagram is offered at this stage purely to illustrate how many things need to be considered when you work out your timetable and the year was chosen as a random example.

The model I have illustrated was drawn up by myself on Microsoft Excel[3]. There is proprietary Project Planning software available that will do the job for you and you pays your money and you takes your choice. Some people simply use a table in Microsoft Word format and some only put in about six key events and 'know' what comes between: I am a compulsive list-maker and need the detail. So long as you (and your Team) know what needs doing and plan for it, the format does not matter. It just has to work.

For a project, where the start-on-site date is not so critical, you can work forwards from 'today' and construct your timetable by adding timescales and ending up with a start-on-site date. If procuring a term contract renewal, however, the start date will depend on the existing provision's end-date and your timetable will need to be worked backwards from then to tell you the latest date by which you need to start the procurement.

This last sentence about term contracts must not be taken lightly. In many professions, such as Housing and Health Care, some provisions or services are statutory duties and some are even vital to life and limb. *These contracts cannot be allowed to lapse!* The importance of forward planning to ensure you start the procurement on time cannot be over-emphasised.

---

3. I use Microsoft systems and so refer to them; this is not to say that other software packages are not equally effective.

The model illustrated in Appendix 2 actually adds the days entered in the time column to the start date and calculates all other dates to suit. Based on a model drawn up by me for term-contract procurements, you can also enter the start-on-site date (i.e. date of end of existing contract) at the bottom and the time in days for each stage of the process and all the other dates will fall into place. It's clever – up to a point – but anyone can do it and, no matter how sophisticated your system, nothing replaces common sense. You must always sanity-check your timetable for gaffes and impossibilities. My example is a copy of a first draft and you will notice it contains Saturdays and Sundays where they are really not practical. Timescales would have to be adjusted to move key dates away from these before this was issued as a final version. Look out for Bank Holidays and weekends and make sure you add in days to make up for delays due to summer and Christmas holidays. If allowing for these takes you beyond your deadline, you may have to reign in some other timescale to make up for lost time. You will see in the illustration that I have built-in contingencies. I like two weeks at key periods but – if required – these can be reduced to nothing.

Always remember when timetabling that not all activities are sequential and many things can overlap and take place at the same time as others. Consider it like a railway line where the process starts on one track and then at a key moment it breaks off onto two or three parallel tracks that all come together towards the end back onto one line. If you study the dates in Appendix 2, you will see how this works in reality and Appendix 7 ('Timeline') illustrates the principle.

Health warning: The Timeline makes it clear that 'Consultation' does not include leaseholder consultation, only 'general' or residents' consultation. This has been done to make the example manageable. In practice you would need to add in at least two 30-day periods of consultation with leaseholders, made a legal requirement by the Commonhold & Leasehold Reform Act of 2002. Leaseholder matters are considered in more detail in Chapter 14, but you need to know here that these two periods are required as well as preparation time for both. This is shown in the sample timetable itself in Appendix 2.

Looking at the timeline for a moment, bearing in mind the time stages are illustrative only and not all tasks are in the list, you will note that tasks have to be completed by the time they are needed (an obvious statement) but the period up to that point can be used to the full. Thus, writing the specification can be started as soon as the intentions are known (as soon as you like, in theory) and be refined throughout the process as the detail becomes clearer until it is time to issue it with the tender documents (ITT stage). The PQQ and associated documents can be started at the same time as the specification, but need to be ready earlier, i.e. when the OJEU is published. And so on.

In simple terms, all the tasks running during one (vertical column) period are running concurrently.

As well as considering the EU timescales laid down for the procedure you are to adopt, and any possible reductions you may wish to take advantage of (see Appendix 7) as well as maybe having to add five-day periods because not *all* your documents are going to be ready to go out (see later where we look specifically at procedures) you will also need to know when to commence pulling together contract documents, Section 20 Notices, Alcatel letters and so on so that they are ready 'for their time'. This may sound a lot and it is, so you may have to wait until later in this book before you feel adequately equipped to do all that.Once you have your timetable, it is your bible. Make sure everyone involved in the project has a copy and that it is frequently monitored and updated where necessary. I often add a column for RAG rating[4] each task, which helps focus attention.

There are various factors you need to take on board when timetabling and we shall look at them now. These are in no particular order of importance – *all* the items are important.

---

4. RAG – Red, Amber Green indicating adherence to schedule. Red items are running late and need urgent attention and pose a threat to the process.

The requirements of your organisation's own bureaucracy

In the course of the following chapters, it will be easy for me to advise you on processes and statutory and advisory timescales. What you will need to do is ensure you have a grasp of the internal machinations of the organisation in which you operate so that you make suitable allowance for any stages of reporting and approval your procurement may have to go through.

Your organisation's bureaucracy will require adherence to certain procurement controls. These will probably include any or all of Contract Standing Orders, a Procurement Code of Practice, Financial Standing Orders, your Procurement Strategy and so on.

In addition, you may well have some form of Gateway process in place that requires all major decision-making to go through a process of hierarchical approval.

Mainly brought about by the recommendations of the Byatt Report[5], a Gateway Process is a form of risk management that requires key decisions to be reached through Reports that:

a) Recommend the decision (or proposal);
b) Have supporting 'concurrent' comments from all key
Departments such as Finance, Legal, Strategic Procurement
and so on;
c) Pass through different tiers of management for scrutiny and
comment; and
d) Finally reach the ultimate (or Executive) decision-maker.

The Report you prepare may be proposing your procurement strategy or it may be recommending Contract Award to a certain provider. The ultimate decision-maker may be the Head of your Department, the Executive Board or the Mayor, all depending on the monetary value of the decision, your organisation's nature and your Scheme of Delegation.

You will need to know what internal processes need to be gone through as part of your tendering and award process and how long each might take. An organisation's requirements may be particular

5. Byatt Report on Local Government Procurement, 2001

to them and you will need to know what they are: if you don't know them, you need to find out, and quick. Someone will know – certainly someone in the Legal section that works on contracts. I can illustrate the importance of this aspect with the following (real) example:

Your Award Report may need to go to Council Cabinet for Approval. But you need to check further than the schedule of Cabinet dates, because Member Support Services may require the final Report to be with them a minimum of two weeks before that date and the draft version of the report with them *three weeks* before. In addition, the Award may be classed (by your organisation) as a 'major' or 'key' decision, in which case it may have to go on a Forward Plan and this notification might be required at least *three months* before the date of the Cabinet meeting. If it is not on the Plan it may not be heard and the decision will not be made. Once the decision has been made it may be subject to what they term 'Call in' for Scrutiny – a method of second-checking the decision. A decision can be called in normally up to five working days after the decision has been made public, which is when the minutes of the meeting at which the decision was made are published. Publishing the minutes can sometimes take up to ten days so the five suddenly becomes 15.

Add that lot up! All for one report at one stage of the process. Gateway processes are only one example of the possible specific stages you will need to cater for in your timetable. I have provided, in Appendix 1, a prompt list or guide to possible internal machinations that you might have to allow for in your procurement. Beware – where you are working, there may be others.

The Nature of the Contract

If a tender is particularly complex or involves bidders carrying out surveys or submitting or analysing large amounts of data or information, it may be prudent to give them longer to submit their returns.

Regardless of what effort a tender submission requires, if you rush the bidders by demanding a return within a period that is unreasonably short, they will see risk in submitting a rushed price and the cost will go up accordingly. If you have to rush or shorten a tender return period, just remember that fact, particularly when you are looking at budgets.

EU Procurement Rules lay down minimum timescales for procurements over the relevant threshold and, if necessary, you can make them longer. Non-EU procurements do not have to comply with these timescales but they are considered to be good practice and, if your CSOs specify minimum timescales, these have to be complied-with. Again, you can make them longer if necessary.

Just because a tender is simple, do not automatically give it a reduced tender period. Only in certain circumstances – and where it is a non-EU tender – can you consider it an option to reduce the tender period. As an example – most Quantity Surveying fees are based upon a percentage of the contract value. If the construction project is straightforward and the requirements of the consultant are limited to QS duties, I see no reason why the return timescale for a fee bid[6] could not be reduced from what might be the EU minimum.

In this context, timescales should only be reduced where:

a) No bidder will be disadvantaged by the reduction.

b) The return date is made very clear to all interested parties.

c) You are clear that there are no complexities or matters of quantity that make the reduced timescale unreasonable.

d) It is *not* in breach of any EU Regulation or internal Code of Practice or Procedure.

e) It is done for good reason.

f) It will not result in you getting poor-quality returns.

g) It will not increase the tendered sums.

---

6. Fee Bid – a tender submission by a consultancy against a brief that gives details of the project.

The Time of Year

When considering timescales and drawing up your timetable, take care to allow for other calendar events that might impede your progress. I have referred to this above, but will re-iterate: do not expect much work to be done between the 20th December and the 6th January (for example) because the Christmas slowdown or shutdown will impact on the progress of your procurement.

You could be bullish and say "if they want the job they will have to do it" and legally you are entitled to say that, but I would be very wary of the outcome. I have mentioned reasonableness – it will keep cropping up – and always let this be your guide. Firms will not give their best offer if some poor underling has been forced to come in to work over his Christmas holiday just to fill in your tender return form.

Also – do not engineer it so that these returns are arriving just as *you* go off for Christmas. If they are going to sit in your office uncared-for, give the bidders the benefit of the extra couple of weeks and ask for the returns back by, say, mid-January.

A similar argument applies to the August summer holiday season, and note that the Christmas and August breaks will probably impact on internal processes as well. Often Cabinets (or other decision-making bodies) will not meet in August or December, and you need to clarify this. If you timetable properly and *start the procurement on time*, these 'stretches' to avoid dead periods will be built in and will not impact adversely on the outcome. In fact – they may even improve it!

Two obvious points – when you make up your timetable, make sure the date you set, for example for return of tenders, is not a Bank Holiday or a Sunday. I know it sounds silly but it has been done! Similarly – do not set 5.00pm on a Friday as a tender return deadline unless you really are going to come in over the weekend to evaluate them. What is the point? Let the bidders – and your tender – benefit from the additional weekend days to get it right and ask for them in by, say, 12.00 noon on the following Monday.

Type and Form of Contract and Contract Period

There are types of contract and there are forms of contract and you will need to know which of what sort you are going to tender as part of your planning process. Do not get them muddled up.

Type of contract

A contract can be one of a range of types, each clearly intended to suit a specific purpose:

- Term contracts will run for a given period and serve an ongoing need or purpose during that time, such as carrying out all running or day-to-day repairs to a portfolio of properties (i.e. an R&M[7] Contract), keeping district heating systems running and maintained, managing a parking system, etc. Providers remain 'on call' and the service is normally provided against tendered prices such as a Schedule of Rates (which is explained elsewhere).
- Projects will have a specific, one-off aim, such as building a community centre, refurbishing a block of flats or replacing the boilers in 225 properties. The contract lays down the work required to complete it.

Projects, in turn, can be minor works or major works, and so the sub-categorisation can go on. In addition to this, contracts can be operated in a number of different ways, depending on how the contract is specified. Specification is covered in detail in Chapter Six so we do not need a great deal of detail here, but a brief overview of the main ways of operating a contract will not go amiss:

- Partnering is a method of contract operation where the client and the provider have shared goals and so share the risks and share the responsibilities to achieve these goals to their mutual advantage (or disadvantage) to ensure a

---

7. R&M – Repairs and Maintenance: a contract for providing a day-to-day responsive repairs service.

successful project. Any financial gains or losses are also
shared.

- Schedule of Rates (SoR) contracts pay the provider the
prices they have tendered for each of the elements of work
or tasks involved in the contract.

- Bill of Quantities (BQ or BOQ) contracts pay the provider
on the basis of a price they have submitted for the whole job
against estimated quantities, but adjusted at final account
stage to take account of changes to these quantities.

- Fully comprehensive contracts pay a provider a fixed
amount per year/quarter/month for a specified level of all-
in service and they win or lose in each period depending on
the level and type of orders raised. Normally used for
maintenance or upkeep contracts.

- Frameworks are another type of contract that you need to
be aware of. They are becoming more and more popular and
are dealt with in detail in Chapter Eight.

So you could have, for example, a Partnered Term Contract
tendered against a Schedule of Rates. That is the *type*. Don't get
confused – you need to be very clear which type you are procuring
if the process is to be a success.

What is the Form?

Form of Contract
'Form of Contract' refers to the chosen document or actual contract
that everyone will eventually sign up to. Some organisations will
have their own, in-house forms of contract for certain circumstances
but it is normal to select from a suite (or set) of standard forms the
one that will best suit the *type* of contract you are tendering.

Contract forms is an enormous topic and out of scope in this
book, but I will explain here sufficient for you to understand the
context of contract form at this stage of the tendering process.

Various forms of contract have been written over the years by various people or bodies to suit specific needs. For example:

- The GC Works suite of contract was written by the Government (GC = Government Conditions) after the last War to provide a legal basis on which to appoint firms to rebuild Britain following the damage caused by the blitz. As a result, the terms and conditions of the GC suite of contracts tend to favour the client.
- The JCT Suite, by comparison, is authored by the Joint Contracts Tribunal[8] and so penned more from the contractors' point of view.
- The NEC 3 Suite[9] – a more commercial version, rapidly gaining favour that advocates more collaborative working
- The PPC and TPC[10] contracts specifically written for partnering project and term contracts.

There are others and it can get confusing[11]. For now you need to know:

- Most organisations have a preferred suite of contracts that they use. Unless you know better – stick with this.
- All organisations will have drawn up their own set of Amendments that will change certain areas of the wording of the contract to better suit their own needs.
- Suites of contracts (by definition) have constituent versions to suit specific applications or *types* of contract.
- Some commercially-available contract forms do not constitute a 'suite'.
- For Partnering, if you wish, you can use any form of contract that suits you and jointly create a Partnering Agreement overlay that stipulates how the Partnering arrangement will

8. JCT – Joint Contracts Tribunal – a body comprising contractors, architects and other construction professionals

9. NEC – New Engineering Contract, drawn up by a panel of engineers for engineering and construction contracts

10. Project Partnering Contract and Term Partnering Contract, created and marketed by solicitors Trowers and Hamlins

11. RICS – The Royal Institute of Chartered Surveyors – is a very good source of information on contract forms.

work. It is intended that the Agreement will replace the underlying contract, and it will do so as long as the partnering arrangement works.

In deciding which form to use, you may 'go with the flow' and use the organisation's accepted preference, and there is often very little wrong with doing that. As you get to know more about other versions, you may find that you prefer different forms and may start to advocate this preference.

Often, a change is good simply to see if a different form *is* more suitable than the one that has been 'used for years'. This is particularly the case as the different suites become updated. From time to time *all* forms are revised and you should ensure you attend seminars that brief you on the changes and always make sure you use the latest version, provided the organisation has updated amendments to match. Bear in mind the 'latest version' may still be several years old and so an organisation's own amendments are always worth reviewing (that means having them reviewed by Legal) to ensure that 'old' contracts are kept up to date with changes in organisational demands and latest contract legislation.

In view of all this, always ensure (to the best of your knowledge) that the Form chosen is the one best suited to your needs and that the amendments are the organisation's latest and fall in line with what you are trying to achieve. Revising amendments will require liaison with your legal department and can be a drawn-out process – hence the need to consider this element in your Procurement Plan. Discussions on this issue should be carried out with your Legal Department early on in the planning stage.

Contract Period
You need to decide for how long the contract is to run. If it is a project – e.g. building a school – that is easy because it will run until the job is finished, although you may have reasons to stipulate a period (e.g. a school extension that needs finishing before the end of the holiday). With term contracts you have much more choice.

It always used to be 'normal' for a term contract to run for 3 years with the option to extend for another two of 12-month periods, so five years in all. In an effort to be more efficient, longer terms came into favour because:

- Procurement costs are incurred less frequently.
- With a longer commitment, contractors can reduce prices (set-up costs and overheads such as vehicles are incurred but offset over a longer period so annual costs are less).
- Bidders will be keener to win a longer-term commitment from a client.
- Reduced (in proportion terms) set-up costs for the client (adapting IT management systems, training staff, etc.).
- Cost-certainty for a longer period for the client leading to better budget management.

For these reasons, longer contract terms have been 'creeping in' and five years plus two (i.e. a possibility of two 12-month extensions) are much more commonplace and periods even longer than that are offered. Nevertheless, long commitments offer less flexibility than you get when the opportunity to re-tender comes up more frequently. For example, a provider may not be good but not so bad that you have grounds to terminate (end) the contract and you are stuck with them and all the contract management issues that go with it. Similarly, your needs might change in the longer term and if you are held in a long-term contractual relationship, the re-negotiation of terms (or prices) may not be as beneficial as a re-tender would have been.

If you are to set up a long-term contract, it had better be a good contract and have either [a] 'break' clauses to allow its termination without any breach on the part of the provider or [b] review periods (i.e. points at which the contract can be stopped or extended) when the contract does not have to be continued.

The decision on how long to make the contract is one for discussion with those best equipped to provide an educated opinion. Primarily, the views of finance, legal and those who are knowledgeable in the field should be sought, along with the opinions of those who will be charged with managing it.

Similarly, in answer to some of the concerns expressed above about lengthy contractual relationships, if the contract is appropriately drawn up, some if not all of the potential hazards can be dealt with at the procurement stage[12]. That is what good procurement is all about – finding solutions that look to the future.

Budget

You must know how the contract is to be funded and that the funds are there. On occasion, it may be that you start a procurement on the basis of funding *becoming* available (e.g. grants) but in such cases you need to know initially from where you *intend* to fund the contract and be sure to make no commitment to any bidder or potential provider or potential bidder until the funding is known to be secured.

Budgets are divided into two large categories – Revenue and Capital. Revenue tends to support ongoing works such as repairs and Capital is normally used for project work where a tangible asset or bankable improvement to an asset results. Generally, it is considered 'good' to capitalise revenue spend and very naughty to spend capital on revenue items. Capital is often available in lumps for strictly prescribed periods of time and, if not spent, is normally lost. Revenue spend is normally planned to come from revenue income over the year-on-year life of a contract and is not so readily 'lost' if there is a delay.

The fine detail of budgeting is beyond our scope here so you must seek the input of your Finance Department at this juncture, but you will need to be aware of the headline considerations so that you can nod in all the right places. For further guidance, if this topic is completely alien to you, courses and books on 'finance for non-financial managers' would strike a helpful chord. We shall, however, look at how the funding needs are determined.

---

12. See Chapter 12 – Risk

What you really need to know about the money is 'how much is there?' as you will certainly need to cut your contract coat to suit your budget cloth. I have never known an occasion when money is not a key factor and you will have to tailor the contract to suit the funds available.

You need to know how much money is available for the life of your contract. If it is a project, this will be a one-off sum (normally capital) to cover conception through to completion. If it is an ongoing term contract, you will need a budget for the procurement process and an annual sum thereafter (normally revenue) to operate the contract for its duration, including any extensions.

Procurement 'people' always try to remove as many uncertainties as possible from every stage of the process (see the chapter on Risk Management) so you should try and confirm exactly what budget you have been given and what it does and does not cover – you do not want any nasty surprises later on. I say 'should' because I recognise it is not always a straightforward matter to confirm a budget, especially when you may be enquiring about it up to eighteen months (or more) ahead of the need.

The budget has to include provision for the procurement process itself as well as the operation of the actual contract. Full-time staff member salaries will normally be covered automatically, but the cost of any staff overtime, additional expertise (such as external consultants) or stakeholder consultation will not be, so their cost has to be considered. You need to ensure that adequate provision for the procurement and contract delivery stages has been made – or else know that it has not, whichever the case may be!

In practice, there will always be a limit on the amount of money available for the different stages in the life of a contract and you, as the procuring officer, will have to manage the process to suit. If the cost-restrictions apply to the contract operation stage you may have to adjust (i.e. rein-in) the specification to ensure that bids come in within the available budget and we look at this aspect in detail in the chapter on specifications. If the money for the tendering exercise itself is limited, you will have to exercise project management skills to avoid an overspend of your own operational budget during this time.

To estimate the required budget, you will need to consider all aspects of the contract's life and allocate them a predicted cost. Things to consider in your cost forecast could include:

Procurement Costs:
- Staff overtime.
- Agency staff support.
- Consultants' fees.
- Provision for consultation – venue hire, catering, etc.
- Trips to view good practice in other organisations.
- Printing costs (eliminated or much reduced if using e-tendering or e-mail).

Contract Operation Costs:
- The cost (probably an estimation) of the (actual) contract including all known extensions.
- Contingency for unforeseen eventualities (mainly with projects – e.g. additional works or bad weather).
- External services (e.g. technical and / or QS[13]) during the operational life of the contract.
- Annual inflation (mainly with term contracts).

There are more, but you will learn of these through the course of this book. Suffice it to say that simply agreeing a cost for 'the contract' is not sufficient. You need to draw up an intelligent estimate of all costs so that you can sit with the purse-holder (or purse-holders – there may well be more than one) and tell them – as near as possible – what resources you will need and how much it will cost to deliver what is wanted, including contingencies.

To carry out this exercise and reach an estimated figure you will need assistance – of that there is no doubt. If you are re-tendering a term contract, you will benefit from the knowledge of those who

---

13. QS – Quantity Surveyor, sometimes called a Cost Consultant – just one example.

manage and work with the current provision, and previous years' out-turns will be a good guide to future costs. For project-type contracts, expert support in the form of a QS and other professionals will be required and – ironically – your cost projections will have to include money for this advice, too!

You will also need to know if there is any provision or leeway for, say, your one-off project extending over more than one financial year or additional sources of funding becoming available (e.g. Government Grants or an underspend elsewhere in your organisation).

There is a greater move towards Whole Life Costing (WLC)[14]as a form of budgeting, especially for construction projects, although the principle can be applied to repairs contracts as well, particularly in the specification of spare parts and replacement units. If WLC is to be applied, the budget calculations are even more complex. Software designed for the task is available, though not always essential, but I would suggest that specialist (or at least experienced) input is a necessary requirement. WLC is beyond the scope of this Chapter (and even this book, really) but it is covered in more detail in Chapter 14.

All of the financial information needs to be collated into a presentable format as it is a fundamental part of the Project Business Case. The level of financial detail you need to include in any report will depend upon the environment in which you operate and approving the budget may be an integral part of the Project Plan or a completely separate approval process. You will need to clarify this at the outset and operate accordingly.

The Project Team

You cannot do it on your own and – as said earlier – you cannot know all about everything. Procurement is very much a team event and your primary role is to lead and drive the process forward.

The make-up of the procurement Project Team is therefore

---

14. See Chapter 13.

vital: it needs to cover all the necessary skills without becoming too cumbersome. To balance these two points, the best way is to have a core team comprising at least one person to cover and lead on each key aspect of the process and appoint to each of them the responsibility to seek additional resources as necessary, but away from your own forum.

The key areas that a project team will need to cover are (in no particular order):

Area	(Primary) Team Member Role
Legal	Ensuring that the process meets legal requirements and providing advice on contracts and amendments
Technical	Advising on the technical and operational aspects of the service or provision being tendered
Quantity Surveying	Assisting primarily with numbers - tender evaluation, pre-tender estimates and so on. A key member of the team at all times.
Finance	Providing budget information and financial advice
Users (residents?)	Focussing on 'what the customer wants' – a key consideration. If a project is being tendered, this may well be 'the client'.
Procurement	Guiding and advising on the procurement process.
Surveying	For construction projects, primarily, to identify and specify the works.

Figure 4.2 – Project Team Roles

Normally, the primary role of each team member will be clear, but in most cases all will have experience of other procurements and be able to make constructive comments on all aspects of the process. Make sure you use this expertise to the full – other people's knowledge and experience are invaluable assets.

The support of an experienced procurement specialist (if there is one) from within the organisation is also recommended, particularly when:

- you are less experienced.
- the procurement is complex.
- the contract is high-profile.
- the contract has elements of high risk.

This person would be one of those whom you would not necessarily require at every meeting, but you will need their advice and input at key points or stages in the process. You will find that the right legal person will know about some aspects of procurement and a procurement specialist will know about some aspects of the legal issues. Together they complement each other and you should have both on your side.

Organising the Team

The demands on each of these areas of the process will vary: it is unlikely you will need a legal expert at every meeting, for example. Schedule your meetings in such a way that only the necessary people are there and ensure you hold whole-team meetings on a regular basis.

A good plan is to agree the frequency of meetings at the very first one you hold, and make sure everyone knows what is expected of them. They *have* to attend or the team cannot function properly and the procurement will be at risk of failure: when tasks are allocated ensure this is minuted and the do-by date noted. If a team member persistently fails, you may have to replace them and in this respect it is always good to secure team members whom you know have the knowledge, the enthusiasm and the reliability to suit your purpose. Do not drag people in 'off the street' to make numbers up – if you do you will regret it later. Trust me.

Each 'arm' of the process may need its own sub-group meetings, led by the relevant Project Group Member, and they will report back to the main group. You would do well to empower them in this, but keep a 'watching brief' to make sure *their* process is driving ahead at the pace which your timetable demands.

'Users' is an odd one. In most Local Authority-type work this will mean residents, and their input is very useful, if not vital, as they are the voice of reality in service delivery. Most often, a couple of residents can be co-opted onto the Project Team. Sourced through a sensible route such as a Tenants' and Residents' Association (TRA), they can be tenants or leaseholders, depending on the nature of the contract, the common practice of your organisation and what the residents want. If it is a borough-wide contract, a TRA will not have the geographical scope to be truly representative and so something akin to a Tenants' Council and/or a Leaseholders' Council may need to be approached for volunteer representatives. If you need both sectors on board, you may have to accommodate more than just two representatives.

These 'users' will help steer the project, but having them on board does not count as full and proper consultation.

Consultation

Consultation is covered in detail in Chapter 14, but I will still consider it briefly here. Whilst you may only need to have two user representatives on your project team you will still probably have to officially consult with all the residents affected by the contract. If it is a block refurbishment, the number of residents may run into hundreds and some of these will almost certainly be leaseholders. As a steer, adopt the consultation process historically adopted by your organisation unless you feel (or know) it is inadequate, when you will have to up the game to suit your situation.

Planning the consultation
Consultation is a specific process, and is considered in some

depth in Chapter 14: you should read that chapter as part of your planning process. I draw your attention to it now, though, because consultation is not an incidental task or occurrence – it requires careful planning and preparation, and adequate time must be allowed within your procurement timetable. For example, the legal requirements attached to leaseholder consultation can add weeks (yes – weeks) to your tendering timescales.

Large meetings such as you might have to hold with the residents of a block of flats will require a suitable venue and ample advance notice, and all of this needs time allowed within the procurement process or timetable.

Chapter 14 explains all of this in detail and you should hot-foot it there now as part of your planning process.

Risk Register

This is a vital tool covered in great and appropriate detail in Chapter 11. Suffice it to say here that every procurement should have a Risk Register set up at the very beginning, and that 'risks' should include threats to the tendering process *and* issues that might arise with the contract itself later on. Many of these can and should be offset or dealt with ('mitigated') at the procurement stage.

But the Risk Register is more versatile than that: as you create and develop it, so it will serve to remind you of things you need to include in the specification and terms and conditions. For this reason alone, maintaining the Register should be an ongoing process.

How long will it take?

The EU stipulates minimum timescales for various and specific parts of the tendering process and these timescales are also considered good practice for non-EU tenders. Under normal circumstances, these add up to between and 55 and 60 days. The Regulations do provide for ways to reduce these constituent periods and you can, in certain, very special circumstances, reduce this total to 25 days. This sounds good but it is not the whole story and we look at this in detail in Chapter 3 on Procurement Routes.

It is sometimes claimed that an EU process takes an excessively long time; the two possible EU timescale requirements given above show that the EU requirements do *not* make a tender process a lengthy operation: it is all the other stuff. In addition to these specified tendering timescales you also have to allow for a whole host of other processes, not least of which are writing the actual specification and evaluating the submissions at pre-selection and tender stages. Do not underestimate the time these activities will take if they are to be done properly.

Additionally, if your contract will impact on leaseholders, you may have a Section 20 consultation process to undertake (see Chapter 14) and this can add a minimum of sixty days (yes – two whole months) to your process.

At the same time, do not think you should always try to reduce tendering times, even where the Regulations allow for it. Rushing the process will leave bidders less time to prepare their submissions and you cannot do this and still expect to get competitive prices for a quality product.

Earlier on I warned against believing those who claim amazingly short periods of time to complete a procurement. If you are undertaking an EU procurement, I would never allow less than a year to do the job properly[15] and achieve a good product at a good price. If the procurement is at all complex, for example a large term contract, I would not be happy with less then 18 months to complete a good process. But that is me.

If you have less time than this, you may be able to get something tendered and secured in the time available, but do not expect it to be the best that you could get at a price that represents best value for (other people's) money. Remember this – no matter who says they can do it.

---

15. This is for a works-type contract; tendering for consultant services can be done in less time.

Summary

- You must have a Procurement Plan.

  It must guide you through the whole process.
- There are many things to consider when putting a Procurement Plan together.
- The timetable is the most important part of a Plan if not of the whole procurement!
- You can work a procurement timetable forwards from today or backwards from the necessary start date.
- Never rush the process and do try to include contingency periods.
- It should be a Team process, so get a good team.

5 // Questionnaires

What are they and why do we have them?

The issue of Questionnaires used to be relatively simple but the 2015 Regulations have complicated the matter considerably and you are advised to look at the Cabinet Office Guidance which can be found at https://www.gov.uk/government/publications/public-contracts-regulations-2015-requirements-on-pre-qualification-questionnaires.

Before I go into much specific detail, I will give you some general points or headlines on what the Regulations state for above- and below-EU threshold tenders.

1. Any tender – works, services or supplies – that has a value below the EU *Services* threshold *has* to use the Open Procedure – so the Questionnaire cannot pre-select (see below).
2. When (1) above applies, Questionnaires can only be used to assess whether a firm meets minimum suitability standards: questions must be pass or fail, not scored.
3. Any tender above the EU Services threshold can use the Open or Restricted (or other) procedure as you choose, enabling pre-selection if you so wish.
4. Any tender above the relevant EU threshold (works, services or supplies) should use the Standard (Cabinet Office) PQQ. This comprises pass/fail 'corporate' questions and 'exemplar' contract-specific questions (see below). This requirement, at time of going to print, is still at its bedding-down stage.
5. Questions can be copied and pasted into your own document or you could – as a corporate decision – adopt the document in its entirety,
6. You do not *have* to copy and paste the questions but the format of your own questions have to mirror the nature of those in the standard document.
7. If you do construct your own questions, it is required that

the questions and criteria laid down are proportionate to the opportunity being tendered.

8. For construction projects it is permissible to use the Industry-standard PAS 91[1] PQQ questions and we shall look at that separately in this chapter.

Bearing these key points in mind, let us now look at some more general issues around Questionnaires.

In earlier chapters you will have seen that any standard procurement involves a stage or element of selection. This is because you only want work carried out by firms who are suitable. Suitability comes under two headings – suitability [a] to work for your organisation and [b] to carry out the services demanded by your contract. Firms' qualifications in these regards are determined through the use of questionnaires and how these questionnaires are deployed in the process depends on the procurement route you use, but there are some general or overall points about them to bear in mind.

• Questionnaires usually comprise two parts:
    ✓ The Corporate Questionnaire or Section
    ✓ A Contract-specific, Technical or Supplementary Questionnaire or (same thing, different names) Section
• The questionnaires are looking at the company itself: namely its capacity, capability, suitability and experience to work for your organisation on the contract being tendered – i.e. assessing *facts*.
• The *tender*, on the other hand, deals solely with the *contract* and looks at a firm's offer or proposals for that contract, normally with regards to quality and cost, on a *competitive* basis.
• You cannot ask a question in the tender documentation that you

---

1. Developed by the British Standards Institute (BSI), the PAS 91 question set has been commissioned by the British Standards Institute and is a recommended common minimum standard for questionnaires for construction procurement. See later in this chapter and view it at: https://www.pas91construction.co.uk/

have already asked at questionnaire stage. More on this later.

- The criteria against which you evaluate or assess the submissions have to be issued along with the questionnaires themselves, including the scoring mechanism (where applicable). People often find this scoring aspect difficult to work round.
- Any responses you give to clarification questions regarding the Questionnaires have to be made known (with the question, of course) to *all* those who have requested the questionnaires.
- Only if you agree that a question is commercially confidential or commercially sensitive can you make the response private to the enquirer. If you do not consider it commercially confidential, you will need to confer with the enquiring applicant and a referral to legal may be required.
  - ✓ Do not respond at all until the matter of confidentiality is agreed.
  - ✓ Do not make any responses public unless the enquirer has agreed or conceded.
  - ✓ If another Applicant makes the same enquiry without requesting confidentiality, you can respond to *that* query openly, including all Applicants
  - ✓ Always – be sure of your ground.
- Always specify the latest date by which clarifications may be submitted and the latest date by which you will respond (normally about 5 days and 3 days before the submission deadline respectively, but this is up to you)
- Firms completing and submitting questionnaires are referred to as Applicants. When they are invited to tender they become Tenderers or bidders. If you are using the Open procedure, they are all bidders.
- Scoring will normally only apply in the contract- specific part of a Questionnaire. The Corporate requirements will generally all have to be met.
- Pre-Qualification Questionnaires must have a 'Pass' mark.
- Business Questionnaires (for Open Procedure) can only have a pass mark for tenders above the EU threshold

- Business Questionnaires for tenders below the EU threshold must comprise pass/fail questions.

Taking on board these overarching points, the matter of when and how to issue the questionnaires needs to be looked at.

When and how do they go out?

When you issue the questionnaire depends on the procurement route selected. Remember – we essentially have the Restricted and the Open Procedures (see Chapter 3) as the basis of all tender processes so shall look at these in some detail.

Restricted Procedure

Under the Restricted procedure, the Questionnaire enables you to see which firms have qualified as most suitable to bid for your contract.

The Restricted procedure is so-called because it allows you, through this Questionnaire process, to restrict in advance the number of firms you ask to tender and so the questionnaire is called a *Pre-Qualification Questionnaire* or PQQ.

Firms will request a PQQ and by returning it they express an interest in tendering for the contract; for this reason, the stage at which this happens is called the Expression of Interest Stage, or EOI. It is also often referred to as the PQQ stage.

Under the Restricted Procedure, firms (under EU legislation) have a minimum of 30 days to request and return the questionnaire. These are calendar days and can be reduced using certain mechanisms (See Appendix 7). The EOI stage can be longer if you so choose.

Firms can request a PQQ as soon as the advert is published, so it has to be complete and ready to go by that date.

The number of firms you invite to bid on the back of the outcome of the questionnaire stage is your decision, provided firms are clearly advised from the outset on what basis the selection will be made. You can restrict the number of firms you ask to bid by:

a) Inviting only the top seven (say) scorers or
b) Inviting any firm that scores above a certain threshold mark, which may or may not be a pass mark (see later).

You can also say you will invite a number between a range (e.g. 'between five and eight'), which allows for several firms achieving scores close to your qualifying mark in the PQQ. To prevent a firm from bidding just because they are one mark adrift may be considered too zealous and would leave you open to a challenge to double-check the scoring. That could end in tears.

There are certain factors, however, that you need to consider.

- Firstly, your organisation's Standing Orders may have requirements in this regard.
- Secondly, you need to be able to demonstrate good competition and sound market testing, so you need a sufficient number of bids returned for evaluation. All tender processes suffer the problem of firms failing to return bids and you should not evaluate if there are less than three[2] returns, so allow for natural 'wastage'.
- Remember - if you invite too many and they all respond, you have a bigger job at evaluation time.
- Bidders will know how many firms are being asked to tender (they are entitled to this information); if there are too many, they may not bother.
- As a guide, a reasonable number to invite ranges from six to eight unless there are reasons for deciding otherwise.

Open Procedure
The Open Procedure does not pre-qualify the number of firms who can tender. For this reason, the Questionnaire is called a *Business Questionnaire* (BQ) and it goes out *with* the tender documents.

---

2. If there are less than three, arguments can ensue regarding levels of competition and demonstration of good market testing.

You therefore need to have the questionnaire *and* all the tender documents ready to go out by the time the first advert appears. This can be a disadvantage of the Open Procedure.

Firms normally (under EU legislation) have 35 days to return the questionnaire and the bid together, albeit as separate sets of documents. Again, there are ways this period can be reduced, and it can be made longer, if you feel it is necessary.

Remember – this is the *Open* procedure; there is no restriction. Any firm who passes the questionnaire stage *must* have their tender evaluated and this can cause logistical nightmares and is an issue discussed in Chapter 3 on procurement routes.

One major consideration you do have with the Open Procedure relates to the contract-specific part of the questionnaire. Because the questionnaire goes out with the tender, you have the option of not having a Supplementary Questionnaire at all but including all the contract-specific questions in the tender documentation and evaluating the responses as part of the bid.

This means you can make issues that would have been questionnaire *facts* – or pass/fail – part of the competitive process instead and this is sometimes considered beneficial. However, it has the disadvantage that you are less-able to be selective about whose tender you have to evaluate and so (many) more bids may need to be processed. Without a Supplementary Questionnaire, the only criterion for suitability to bid would be that firms would have to meet the requirements of the corporate Questionnaire. Normally, many firms would do that and you will have made no distinction on the basis of experience or capability.

As I say, you do have the choice, but I would advise that a Supplementary Questionnaire forms part of the BQ to help control the number (and quality) of the bids you need to evaluate.

Pass Marks for Business Questionnaires
If the tender is below the EU Services threshold, Questionnaire questions have to be scored on a pass/fail basis, meaning that a weakness in one area cannot be compensated for by a good score in

another part. If the tender is above the EU threshold, you can score questions and will need to declare a pass mark whereby it is decided whether an applicant is eligible to have their bid considered or not.

Of course, all information relating to the question of pass marks and eligibility must be given in the documentation – whether it's an Open or a Restricted procedure.

Creating the Questionnaires

I shall first alert you to and remind you of some new requirements brought in by the 2015 Regulations.

For any tender below the EU *Services* threshold:
- You cannot issue a PQQ (i.e. you must use the Open Procedure)
- You can still issue a capability Questionnaire (Business Questionnaire)
- Questions must be:
    - ✓ Relevant to the subject matter of the contract
    - ✓ Proportionate
    - ✓ Only used to assess whether minimum standards are met
    - ✓ Be Pass/fail – not scored.
    - ✓ Be aligned with the Standard PQQ

For tenders above the relevant threshold:
- You can pre-select with a PQQ – if you wish or you can still 'go Open'
- The Questionnaire must comply with the statutory guidance issued under Regulation 107(2) of the 2015 Regs (you need to look this up) which states that...
- The Cabinet Office Standard PQQ must be used for the Corporate Section ("should be adopted")[3]
- Contract-standard questions in the Standard Questionnaire must be used, so far as is possible – they may be tweaked or

---

3. I have assumed that 'PQQ' means 'Questionnaire'. This is one of the vagaries, so adopting the Standard form for both avoids the risk of challenge on this technicality.

omitted to suit specific instances
- Non-relevant questions must be omitted
- The Standard PQQ comprises core questions on:
  - ✓ The supplier or applicant's details
  - ✓ Confirmation that they are eligible to bid with relation to the grounds for exclusion covering fraud, embezzlement, etc. [4]
  - ✓ Economic and financial standing
  - ✓ Technical and professional ability – previous experience
  - ✓ Technical and professional ability – with respect to the proposed contract
  - ✓ Insurance
  - ✓ Equalities
  - ✓ Environmental Management
  - ✓ Health and Safety

More Information

The regulatory requirements regarding Questionnaires are complex and new and still being 'bottomed out'. For this reason, you need to read the Guidance issued periodically by the Crown Commissioning Service. This can be downloaded at:

https://www.gov.uk/transposing-eu-procurement-directives

You are also strongly advised to refer to Procurement Policy Notes issued by the Cabinet Office at:

https://www.gov.uk/government/collections/procurement-policy-notes

I also strongly recommend the Cabinet Office Guidance on

---

4. Rejection of an Applicant on these particular grounds has now been granted full legal support under the powers of The Public Procurement (Miscellaneous Amendments) Regulations 2011.

Questionnaires at:

https://www.gov.uk/government/publications/public-
contracts-regulations-2015-requirements-on-pre-qualification-
questionnaires

Here you will find a copy of the Standard PQQ. You should download
this for cross-reference as you consider this chapter.

That's a lot of links here and the reason I advocate this approach,
rather than iterate the various guidances in this text, is because the
guidance is constantly changing at the moment: new Regulations
require interpretation and testing and the 2015 Regulations still
(almost) have the wrapping on: there is (at the time of going to
print) a lot of uncertainty about what the Regulations actually
require and how they will be interpreted and we may have to
await the outcome of a few test cases before we have any degree of
certainty.

My suggestion is that, as new guidance is published, you print it
out and tuck it into this chapter as an update. Highlighting in this
book what has changed will also be a help.

Getting it together

Bearing the above points in mind, and as a reminder, there are two
distinct parts to the questionnaires:

- The Corporate Part
- A Contract-specific, Technical or Supplementary Part
  (remember – same thing, different names)

Corporate Section

Whilst the Corporate Section will most likely be managed by a
central or strategic procurement unit, the approach towards it
is relevant and you may even have to manage all or part of the
process, depending on the ways of the organisation in which you
operate. We shall therefore look at it in a little detail.

The Corporate Section must now follow the Cabinet Office standard format. You may tweak questions and could change the format if you wish (e-tendering systems may mean you will have to) but why bother if you don't need to?

The Corporate Section is almost always Pass/Fail – a firm is suitable to bid (meets your minimum requirements) or it is not. By virtue of its nature, there may well be some flexibility with the financial assessment (see Chapter 16) but generally that is all. The responses to these corporate questions are often evaluated by a central team against standard, published criteria (that is, they must be issued with the questionnaire, at least).

Bear in mind that legal requirements (for Equalities and Health & Safety, for example) will often depend on the size of the company and their evaluation will probably (with a bit of luck) be out of your hands anyway and managed by specialists. This is certainly the better option. Suffice it to say, if a firm does not 'pass' the criteria they are not deemed fit to work for the organisation and so cannot be asked to bid (Restricted) or have their bid considered (Open).

Many firms, especially smaller ones, struggle with Questionnaires so it is a good idea to have clear and simple guidance notes that explain the Corporate Questionnaire and the Evaluation Criteria, laying down any terms and conditions attached to submitting an application and advising on relevant legislation, areas of tolerance, etc. The Cabinet Office does not lay down evaluation criteria but the Regulations require you to make these clear at time of issue, so you will need to construct them. The corporate requirements may be outside of your remit: you will need to go with your organisation's flow on this one unless you have particularly strong views on the matter. If you have to set them down you will need to liaise with the relevant areas within your organisation to determine minimum requirements against which submissions may be assessed and set these out clearly within the documentation.

How you set this out is up to you – just make it neat and simple, either lay down the criteria with the question or set them out in a separate, explanatory document. Try and do this in such a way that you end up with a standard document applicable to all procurements and do not have to repeat the exercise. It is worth taking time on this.

It is possible to ease a firm's workload when submitting a Questionnaire in any one of the following ways:

- Accounts can be downloaded from the records of a Credit Rating Agency if your organisation subscribes to this service. Do *not* simply assess financial standing on the basis of an agency's credit rating: different agencies may have different ratings and the use of any rating for this purpose is against the Regulations
- Some e-tendering systems allow firms to upload a Corporate PQQ submission 'master' and draw it down for repeated Applications
- The Regulations also allow firms to self-certify: you state your requirements and they can confirm (perhaps in a simple tick-box) that they meet the requirement. You then accept this and only seek proof if and when you get to the stage of award. This concession may be fraught with potential problems but the Regulations encourage it and documentation is scheduled to be published to support this approach.

Normally, as I have said, the Corporate Questionnaire is pass/fail. However, if a firm submits – say – an insurance policy document that is out of date, it is not untoward to call them and ask them to submit the current document. There does, of course, have to be a limit to the level of tolerance and assistance you show in order to ensure that the procurement process remains 'good'; there are no written rules on this, but some useful rules of thumb are:

- Adopt a 'reasonable' approach – try and distinguish (say) between a firm mistakenly attaching an out-of-date document and simply not being able to meet your requirements.
- Your approach in this regard must be applied in an equitable manner to all applicants.

- Avoid clarifying or assisting so much that you have ended up telling firms what to put in their answers. To be honest, the published criteria have already done that for them: if you have to be even more 'helpful' then maybe they are not the right firm for you.
- When requesting additional information or clarification, set specific deadlines that cannot be breached.

These are good rules of thumb that apply equally to both the Corporate Questionnaire *and* the Supplementary Questionnaire. We shall look at the Supplementary Questionnaire now.

Supplementary Section
As stated above, this is the contract-specific Section that should demonstrate a company's capacity and capability to provide the actual service you require. It is *not* a bid. To my mind, the Standard PQQ is weak on contract-specific matters, but there is not much we can about that other than to 'beef up' the quality section of the tender itself.

The Questionnaire asks for details of previous, similar contracts. It does *not* specify how you evaluate or score submissions so I offer some guidance on that here as the Regulations also require you to make your scoring model completely transparent at time of issue.

You can be quite specific on how closely the reference contracts match the one you are tendering and you should ensure the details submitted include contact details that will enable you to seek references. You can reserve the right to fail an applicant/bidder if you deem the reference contracts do not adequately reflect the contract you are tendering or if the references do not support or demonstrate the applicant/bidder's ability or capacity to perform the contract.

If the practice is accepted to broaden the scope of the Contract-Specific section, then I refer you to Chapter 7 on Evaluation Models. Whilst the Questionnaire has to ask about previous experience and current capacity and capability, the method for setting up the questions, weighting them and scoring the submissions explained in Chapter 7 all apply and can be read in the context of the Supplementary Questionnaire.

PAS 91

PAS 91 is a standard drawn up by the British Standards Institute (BSI) and provides guidance on selection questions for questionnaires. Its own introduction states *"BSI PAS 91 is a publicly available specification (PAS) that sets out the content, format and use of questions that are widely applicable to prequalification for construction tendering."*[5] This is a very comprehensive document and the Cabinet Office Guidance allows its use instead of or alongside its own Standard PQQ. It is accessible at: https://www.pas91construction.co.uk/

Setting Pass Marks for PQQs

In the Restricted Procedure, you can simply say the seven highest scoring applicants will be invited to tender. This is fine, so long as all top seven score enough to prove they are suitable as providers. This begs the question "What score means 'suitable'"? In other words, what is a pass?

Additionally, if you select the top seven and one drops out you can substitute another applicant to bid, provided they are suitable – in other words so long as the eighth passed. Again, you will need a pass mark, where 'pass' means 'good enough to bid'.

Do not stab at it. Simply saying '50% will be a pass' may let in rubbish, whilst saying '70% will be a pass' may eliminate good companies. To determine a meaningful pass mark you need to go into each question and ask yourself, considering the context of the matter being evaluated, "what score should a competent company achieve *as a minimum* or what minimum score would be *acceptable* from a company on this question?" It will not be top marks for each

5. PAS 91:2013, page v.

question, nor perhaps even half marks for each question. Remember – look at each question in turn, and consider it *on its own merits*.

You can then tally up the notional weighted minimum or acceptable scores for all the questions and arrive at a pass mark. This is what you declare to applicants.

When you have done all this, get some critical friends to have a go at some mock submissions and evaluations and see how the model works. Remember to allow time for all of this in your timetable.

Evaluating Questionnaires for the Open Procedure

We have already covered the way Open Procedure Questionnaires must be scored, but we need to consider how to minimise the task, especially as it is possible to receive a multitude of tender submissions in response to an Open procedure advert. There are a couple of ways the Regulations assist in this:

1. Of particular use when the tender evaluation is a relatively easy process, you can evaluate that element of the submission first and then consider the Questionnaire element of the 'winning' bid'. If they pass the Questionnaire then you have the winning bidder. If they do not pass the Questionnaire stage, then you need to progress onto the second-in-line 'winning' bid, and so on. This can save time and effort.

2. Bidders may 'self-certify' in that they can confirm (usually via a tick box) that they have, for example, the required levels of insurance or the necessary accreditations and you only need check these if they accede to the position of winning bidder. Again, if it transpires they do not meet the requirements, then you cascade down to the next in line. This will require contacting them to obtain the information but it does save you ploughing through unnecessary documentation from each bidder and it saves them attaching it.

You may choose to adopt either of these approaches – but remember to make the process you are using clear in the documentation.

Additionally, you can take advantage of the second option even in the Restricted Procedure when you use a PQQ – but you would have to check the validity of any self-certification prior to inviting an Applicant to tender.

References

References always promote discussion, mainly because they are important but also because there are no real rules. I include the issue here because whether or not to include references at the Questionnaire stage needs to be considered. The answer is – you can if you want to. My view is that you should, but I would recommend not taking up references until you are considering a company (or the final three) for contract award. Why seek more references than you need? Just remember to explain to Applicants what your process will be.

You can seek references in several ways:

- If you are asking for details of other contracts similar to the one you are tendering, make sure you ask for a client name and contact details and advise the bidder that you reserve the right to approach any of those listed for a reference.
- You can ask for a list of six referees and reserve the right to approach any or all of them for a reference prior to any contract award.

I would suggest you give the Applicants a table to complete so that you know you are getting the all information you require in a legible format.

Whilst such information can be sought as part of the tender process, it is my view that it is neater to secure this information as part of the Questionnaire process rather than at the competitive bid stage. It *is*, effectively, looking back not forwards and also: it gets it out of the way.

If you do seek references at the tender stage, you need to be careful to seek opinions on the bidder with a view to them performing *your* contract, rather than simply asking how they did before. It is all a question of phraseology.

'The Third Way': This is a little tip that you may choose to employ. Because seeking references can be a time-consuming process, this 'trick' can ease the burden. You can include the reference request form with the documentation and have the *Applicants* (or bidders if at tender stage) seek the references themselves and have them sent straight to you. Make it a condition that the submission will not be considered unless you have received the required number of references for each bidder and leave the toil to them.

Finally, how to assess references. Always make it clear that contract award will be subject to the bidder securing satisfactory references, but you will need to define 'satisfactory'. To do this you will need to concoct a paragraph that includes phrases such as "...in the opinion of the Evaluation Panel..." and so on, to give you the leeway to make a subjective decision (because it can often be no other).

As always, there is another way, although many people advocate against it. If you have any doubts about it but fancy a go, you can make the referencing process clear in the earliest tender documents and you will be covered because bidders will have ample time to object if they feel there is a fault in the proposal[6]. This will mitigate any risk that might be perceived in using this 'scored' reference system.

I have used this method and it does work, so I offer it here purely as an option in your armoury for you to consider: You ask referees to score aspects of the bidders' performance out of 5 (5 is super, 1 is lousy, etc.); you state that any single score of 1 will rule a bidder out and they must score – say – an average of 3 out of 5 overall.

6. I refer you to the 30-day time limit for raising challenges under EU Law, which would be considered adequate and good practice in a non-EU procurement. If a bidder continues to tender, their acceptance of the proposal would be implied.

Assessing the references becomes easy and referees like it because it is quick and simple.

Health warning: do *not* use this as part of a tender evaluation score: references are purely references.

Issuing the Questionnaire

The way in which the questionnaire is issued will depend on what tendering processes your organisation has in place. We will look at full-blown e-tendering in Chapter 15 but, where e-tendering is not in place, most Questionnaires now are requested by e-mail (the advert will advise how firms are to request one) and the Questionnaire is sent out as an attachment as soon as the request is received. The questionnaire therefore has to be ready to go as soon as the first advert is published.

Some organisations may persist in sending out documents by surface mail, but this is a diminishing idiosyncrasy and soon to be banned by Government legislation (2018, in fact, for most of us). E-mail used to be a common alternative but full-blown e-tendering is rapidly becoming the norm. Organisations are securing their own provision and those that aren't make full use of publicly-available and approved e-tendering portals.

If the process is not fully e-enabled, submission of the completed questionnaire can be by e-mail attachment or hard copy posted or couriered (best and safest). Remember – this is not a competitive stage as such, so there is no need for sealed submissions, ceremoniously opened when the deadline passes. E-mailed responses can be downloaded and stored – whatever you like – so long as they are safe and ready for the evaluation process to begin.

It is intelligent to keep an e-mail distribution list of all firms who request a Questionnaire. Remember – any clarifications have to be issued to all. As you get a request, add the address to your list and then contacting all Applicants becomes easy.

Two tips about e-mail processing:

a) Paste the distribution list into the address bar *after* you have completed the e-mail message and completed any attachments. This avoids accidently sending anything before it is finished
b) Do not use the 'To' address line. Use the 'Bcc' address line then no applicant – rightly – will know who else is involved in the process.

These two tips apply equally to the tender process as well as the questionnaire stage.

Do not start evaluating before the deadline. It is quite in order for a firm to contact you because they wish to add or change something in their submission. If you have started work on it, this is more difficult, and could lead to accusations that you have provided advice on their responses. Leave submissions alone until they are rightfully yours – i.e. after the submission deadline has passed.

Summary

- Questionnaires assess the suitability, capability and capacity of *firms* wishing to tender
- You must use the Cabinet Office Standard PQQ for tenders above the threshold and
- Translate this into a format suitable for any procurement below the threshold
- PAS 91 may be used as a model for construction procurements
- It is not a competitive process, except that you may have set criteria or rules regarding pass marks and bidder selection
- Requirements must be proportionate – this is particularly sensitive for below-threshold procurements
- You must make clear *how* you are going to assess responses
- The Questionnaire stage is important as this is what secures good firms for you to ask to bid.
- You can be tolerant in seeking clarification of responses, but you must be fair to all parties and not over-helpful
- You may be challenged on the outcome of your evaluation, so you *must* make sure it is reasonable, equitable, logical... and *correct*.

6 // Writing the Specification

What is a specification?

The specification of any contract is *the* tool within the procurement process that determines what you will get from the contract: it lays down details of the 'product' required and its quality. All the other bits in a set of contract documents are primarily designed to ensure you do get what you specify and identify what you can about it do if you don't. For this reason, a specification should never be 'run of the mill' or simply taken from another document. It should, instead, be drawn up for the purpose, using thought and deliberation to ensure that maximum value and benefit are gained from the process.

It is against the specification that instructions will be issued, work monitored, and a contractor's performance judged. For this reason, the process of contract *management* can never be separated from the process of contract *procurement* and the writing of the specification in particular. Figure 6.1 (overleaf) illustrates the inextricable link between the tendering process and the management or operational stage of a contract.

Point to Note
Whilst we shall be concentrating on writing works-type specifications, it is worth noting the additional leeway that the new Regulations have allowed in defining products and materials in any tender. It is now even encouraged to specify quality and other accreditation marks such as Red Tractor and Fair Trade (for food products), provided you will allow other, equivalent marks as well (the accreditations would have to be checked and approved by you).

It is also in order to specify requirements regarding the manufacturing process of goods and products to be used in the execution of your contract. For example, clothing might have to be manufactured in workshops that have a proven ethical code of conduct towards its workers; building materials might have to have a low carbon footprint."

The only thing is, you can *only* specify such requirements for products to be used in fulfilling *your* contract: you cannot insist on it being a standard the provider meets in all of their contracts.

There is more on this under 'Sustainability' in Chapter 13.

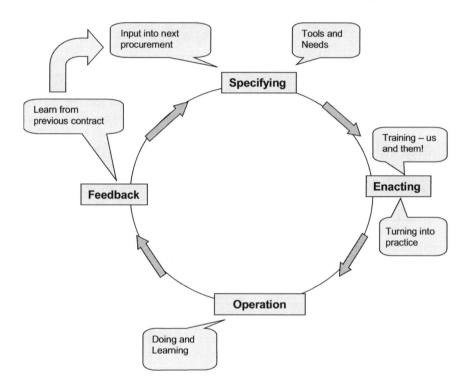

Figure 6.1 - The Contract Management and Procurement Cycle

Looking at Figure 6.1 above, the table in Figure 6.2 below will explain each part of the cycle and how they interlink. Working clockwise:

Stage	What it means
Specifying stage	Writing the specification
Tools and needs:	The specification determines what must be provided and how it is to be done.
Enacting stage	Turning the written word into the actual delivery of the contract.

Training – us and them:	All staff involved in the management of the contract should be trained in the contract's operation before it commences. This is particularly important for term contracts but may be less necessary for projects. Similarly, the providers (contractors) will probably benefit from a two-hour induction, as those supervising the contract will not have been involved in the tendering process and it will be new to them; this should be the start of working with them.
Turning into practice	The process of getting the contract working and turning the theory of the specification and terms and conditions into a working reality. Includes the award stage, pre-contract meetings, clarifications, introductions, lead-in times, etc.
Operation stage	The contract actually working ('on site')
Doing and learning	'Doing' is the service being delivered and the client learning from this, taking on board what does and does not work and storing this information for feedback into the next procurement. (Jargon – "lessons learned").
Feedback stage	Relaying the experience gained in the operation of the previous (i.e. ongoing) contract
Learn from previous contract	The information and experiences gained from the contract's operation are fed back into the new procurement process. Change weaknesses, use and build on strengths.
Input into next procurement	Those with the contract management experience should be part of the Procurement Project Team and contribute their operational knowledge and experience. The specifier learns from this 'knowledge bank'.
Specifying stage (Again)	The new specification now incorporates and builds on those parts of the previous specification that worked well and changes those areas where it was considered the specification 'lacked' what was needed.

Figure 6.2 – The Procurement Cycle Stages Explained

When writing a specification, the following headline issues should be considered:

- Budget
- Value for Money
- The need
- Meeting the need
- Added Value
- Contract Manageability

Looking at each of these in turn:

Specification Writing – Budget

At the planning stage you will have determined your budget for the operational stage of the contract and you will need to keep this limitation in mind when writing your specification. The aphorism 'avoid champagne taste and beer money' is a good edict at this stage. Your QS will help you to ensure that you remain within your budget by providing Pre-tender Estimates (PTEs) of the cost of your proposed specification. If you are likely to run over budget, you can either seek more money to meet the additional cost or you can reduce the cost of the contract by the use of various methods, including:

- reducing the specification – e.g. project scope, material types, quantities, etc.
- making contract timescales less demanding, so reducing the provider's overheads.[1]
- setting less-essential items aside within the bid: they can become 'provisional items' which, whilst priced within the tender, can be included in the works only if and when the budget permits.

There are other, more 'elaborate' methods of reducing costs, such as combining schemes to increase the size of the contract (which could include the use of Lots and Packages or piggybacking an existing procurement, which will also save on tendering costs). Using existing

---

1. In a project, this can mean more reasonable completion dates; in a term contract, easier response times – if possible – will also help with this.

Frameworks can also be considered: these will not guarantee lower tender sums but may help reduce procurement costs.

Writing process or workmanship specifications is an area that can quickly get out of hand and escalate costs so care needs to be taken to ensure that the aims of the project are kept in focus. As an example: when specifying the external redecoration of a building with wooden windows, the following questions will arise when considering what needs doing:

- Do you burn off *all* the existing paint down to bare wood or do you allow a burn-off of, say, 20% of the windows?
- If the latter, how will you measure this on site?
- Burning off using gas brings a risk of fire but electric heat-strippers are very slow
- Paint stripper will also obviate the fire risk but takes a very long time and can be very expensive
- Do you specify repairs to frames in timber or in a modern, resin-based alternative?
- Should all windows be left fully working (this may include replacement of sash weights, for example)?
- Do you treat bare wood with preservative or go straight to primer?
- Do you use oil-based or water-based paint? (Water based is quicker to apply and quicker to overcoat, but is it a better-performing product?)
- Is it worth specifying Microporous paint?
- How many undercoats and topcoats are to be applied?
- Do you insist on different shades of paint at every coat so as to ease identification of completed stages by the Clerk of Works on site?
- Whose products will you propose and so what level of paint quality?
- Are you seeking, say, a 10-year or a 5-year projected life?

This is a simple example, and the answers to these questions are beyond the scope of this book. However, whatever the project, similar questions will arise all the way through a specification process and they will need to be considered and then addressed in a way that can be justified in terms of final product quality, cost and time.

Lots and Packaging

Some contracts lend themselves to Packaging or Division into Lots. The principle involves dividing a contract up into composite parts and allowing firms to bid for one, some or all of these parts whilst offering the buyer discounts for multiple-Lot awards. I mentioned earlier piggybacking one contract with another to gain economies of scale and reduced tender costs: the component projects could each be a 'Lot' within one tendered contract.

Whilst not always beneficial, and requiring a more complex evaluation model, packaging can be a useful procurement tool in the right circumstances. The potential advantages, in addition to reduced tendering costs, include increased market opportunities, more favourable bids and flexibility in the procurement and award processes.

With service or maintenance contracts (e.g. property maintenance), laying down geographical limitations on the area to be covered could make it easier for a company to operate, enabling them to provide a higher-quality service at a lower rate, so division into Packages or Lots by area could achieve this. Similarly, a clutch of relatively small building projects could all be tendered as different Lots within one contract.

When a tender is divided up in this way, larger firms can bid for some or all of the Lots whilst smaller firms have the opportunity to tender for perhaps just one of them, so opening up the market to SMEs[2], in fact a package could be set aside as individual from all the others specifically to encourage SME bids. The use of Lots or Packages is therefore often a useful option, assisting with budget control and tendering costs and is always worth considering.

2. Small and medium-sized enterprises – normally defined as having up to 250 employees. Local Authorities have a duty to encourage and support such businesses where and when they (legally) can. This comes under the umbrella of economic sustainability.

Provisional sums

Provisional sums can be set by the client within the pricing document for works where costs cannot be predicted at time of tender or (paradoxically) where specific material costs are already known by the client. As items within a provisional sum are already costed by the client they do not require a price bid from the tenderers, they are just included in the bottom-line cost.

Contingency Sum

It is always advisable to have a Contingency Sum within the budget to cover any unforeseen items or problems that may arise. By their nature, these items cannot be predicted so the size of the contingency is a matter of conjecture, but up to 10% is considered reasonable, depending on the complexity (and thereby the unpredictability) of the works. A risk assessment will help determine the sum best set aside for contingencies.

Always ensure the Pre-tender Estimate (PTE) for your specification – including provisional and contingency sums – is within the allocated budget: do not tender if you expect bids to exceed the amount of money you have available (in a Local Authority environment, to do so is not legal).

In project and repairs contracts, if the tenders still come in above your budget, there are methods you can use to bring the costs of the project back on track. You can re-tender with items removed (least desirable) or you can issue tenderers with a 'Bill of Reductions' – a tender device that asks them to submit a price against those items you wish to remove from the original specification. You can then deduct the submitted sums from the original bids and see who now offers the best tender price.

You need to be sure, whatever route you choose, that any such actions are executed in a way that avoids challenge from bidders, and them claiming that the process has been carried out unfairly. You should always ask *all* bidders to partake in a Bill of Reductions process – not just a chosen few. The logic of this is that, apart from fairness, you do not know how the bids – and their distribution of profit – have been composed and the impact of removing items from a pricing document can radically change the initial order of a set of submissions. For the same reason, it is dangerous to simply remove items from submitted tenders at time of evaluation. You may take out a chunk of a bidder's profit and make the remaining works impossible for them to viably undertake.

Where you know you are 'pushing' your budget limit, the solution normally used is the inclusion of removable items that are priced separately and identified as removable within the documents. This way, there are no arguments.

There are some methods of specifying works that lend themselves more readily to cost control than others, and we shall look at these later on (see Specification Writing – Meeting the Need, below).

Always bear in mind, however, that any additional, cost-reducing process will add time onto the tendering period and your timetable must allow for any such hold-ups. Remember timetable contingencies?

It is sometimes thought possible, under such circumstances, to omit items from the actual contract once it is on site. No - you cannot legitimately award a contract (i.e. make a financial commitment) if you know in advance that it is above your budget. In reality, in most (and hopefully all) organisations, this would not be possible. The only time such an action might occur is if the tendered sum is exceeded for any reason during the course of the works, when there are two possible courses of action:

a. Lower-priority works can be omitted to bring the project back on financial track, or

b. More money can be sought. Authority for this additional cash must be sought from signatories that have a level of financial authority for the new total value of the contract, which means the original sum plus the additional funds.

If the extra costs are down to additional work being uncovered or more works are (legitimately) required as part of the original project, a *variation* to the contract needs to be drawn up to account for them.

You need to beware, though, that if additional works are *desired* (as opposed to additional works being *necessary* to overcome unforeseen problems) you may be open to a challenge from the other original bidders that the contract finally undertaken was not the one that was actually tendered. You may scoff and say you could disprove such a challenge, but such success does not necessarily wipe out the costs and delays such a challenge may incur.

To avoid this, always ensure that any such variations are within the 'scope' of the original contract. If the contract is for cutting grass and you want a few bulbs planted, this may well be considered to be in scope. If you want the grass-cutters to clean the windows as well, this would not be in scope. There are no hard rules – judgement is always required – but essentially you need to go back to the original advert for the contract and decide if the extras lie within the scope as advertised.

Specification Writing – Value for Money

The term 'Best Value' (BV) usually implies a procurement carried out under the Four 'C's and was used to imply tendering in a manner that would have the best outcome for the client or end-user. It is not within the scope of this chapter to provide information on the process of Best Value in any great detail, but it has been touched on – as has its most recent persona – in the introduction to this book (which I hope you have read, by the way).

For the purposes of this book, best value in its more 'modern' context[3] means that a contract achieves what is best for the organisation and its customers in terms of value for money. There is an obvious argument that says that lowest price must be the best value but of course this is not always the case – for example, rubbish is cheap.

Evaluating bids on the basis of lowest price does have it place, however, and is generally used where the 'product' can be rigorously defined. Let's look at two examples.

---

3. Best Value – See also Chapter 16

Example 1 – If you place an order for 12 fridges to equip a new sheltered housing unit you have just built, you can specify precisely the make and model, delivery requirements and installation (left working with all packaging removed from site). Easy – lowest price will do, just as when you go out to buy a specific model of a new fridge.

Example 2 – In construction, there is a very detailed client specification, a mass of rigorous Building Regulations, a Clerk of Works on site to monitor quality, a surveyor to check work on handover and a QS who will not authorise any payment until all quality and quantity issues have been resolved. This is again an environment that closely controls what you get for your money and lowest price can often be used for 'standard' (or 'run of the mill') building projects.

In both of these examples, any legitimate supplier should be able to provide the 'product' specified and proper contract management (even if 'contract management' simply comprises the process of receiving and signing off the delivery) will ensure that it is. The problem arises when there are potential areas of flexibility or tolerance in the product's specification or the providers' proposed methods of delivery, and the quoted price cannot be the only consideration. The areas where questions arise about how well something is delivered are known as areas of quality and, when they apply (which is in most cases), they need to be considered as part of the bid, alongside the tendered cost.

The preferred term for this within procurement circles is Most Economically Advantageous Tender – or *MEAT*[4] for short. This may mean that the bid accepted is not necessarily the cheapest but is the one considered to afford the maximum benefit to the client and its customers, in terms of the balance of price and quality, within the available budget.

---

4. The 2015 Regulations do not recognise lowest price, only citing MEAT as an eligible evaluation model. However, it does recognise that seeking the lowest price may represent MEAT: same thing with a little syntax thrown in for regulatory good measure.

At specification stage, it is necessary to be absolutely clear what the quality aspects are and what level of quality you are seeking, making sure all the while that the budget will sustain it. In a services contract (for example, responsive heating repairs), value and quality may be assessed in terms of labour skills, management processes, customer care processes, stocks of spare parts and so on. The order of priority of these aspects will be determined through the consultation process and their value *relative to the price* you pay will be judged within the evaluation model you create.

We shall look at evaluation models later on, but always be careful when consulting with end-users on quality and ensure that you manage their expectations, as the budget will be limited. For this reason, prioritising quality aspects gives you the flexibility to trim their aspirations as necessary in order to meet financial constraints.

In construction projects, the question of quality of finish is largely client-driven although as explained, within the specification, much of the build quality will have been determined by statutory requirements such as are laid down in the Building Regulations. Remember – you can specify a higher standard than the Regulations require but never a lower one.

Quality on a construction project can extend beyond the work and materials. For example, you could take into account the contractors' approach to a project's impact on the immediate environment. Accreditation by a body such as The Considerate Constructor Scheme[5] is one way of assessing such a criterion but you could also specifically consider working times, vehicle movements and other 'quality' aspects relating to the location of the site within the quality element of your evaluation model, using method statements to assess bidders' proposals.

These types of considerations come under the heading of 'Sustainability' and are becoming matters of prime importance. Local Authorities and other public bodies, in their hallowed

5. The Considerate Constructor Scheme aims to encourage building and civil engineering contractors to carry out their operations in a safe and considerate manner. See link for details: http://www.considerateconstructorsscheme.org.uk/
Please note: other, similar schemes exist.

position, have to be at the vanguard of such considerations. We have devoted Chapter Thirteen to sustainability, but further research by you will be required – it is a big topic in its own right (understatement) – and your organisation will have its own priorities under this banner.

Your QS will help you manage client[6] expectations in a project. Remember – it is not for you to tell the client what quality of carpet they should have in the reception area. If they want cashmere and can afford it, it is for you to specify it. However, if the budget will not run to it, it is for you to advise them that something is going to have to be compromised if the tender is to succeed. It may be that you ask bidders to price more than one option – you never know, you may be able to afford the better option – but good procurement has to maximise opportunity without knowingly tendering something that will not achieve its objectives.

You are referred to the section on Budgets, above.

Specification Writing – Defining the Need

The need is essentially 'what the customer wants' and has to be clearly defined and quite detailed. Depending on the nature of the project, the needs 'list' may be drawn up by an individual, a panel or through a consultation process that takes on board the views of many stakeholders – whichever route is most appropriate to the project.

Mission Creep

A construction or repairs project will often start off with quite a focussed aim – e.g. "external decorations to a building" – but will then gradually expand as the opportunity is taken to address other issues that prevail. This is especially the case when the project includes the erection of scaffolding, for example: work to the roof, new windows and replacement rainwater goods will

---

6. 'Client' could be your managers, another Business Unit or any organisation that employs you. They are the ones responsible for the contract.

soon start being added to the list. This is called 'Mission Creep'.

It is a fact that, if you have a consultation meeting with stakeholders to discuss the specification for a project like the example above, you will need to manage expectations in order that the 'needs' list remains in touch with the reality of the budget and does not unwittingly become a wholly unattainable 'wants' list.

It is necessary to ensure that the needs list is then priced and prioritised, so that the project can be achieved within budget (this is part of the Pre-Tender Estimate process). The resulting prioritised list will identify the items that have to be done, those that might be carried out depending on the value of the returned bids and those that will definitely have to wait for another time.

Once the needs list is finalised, it has to be written up into the form of a proper specification, suitable for a bidder to price.

A *term* contract specification (such as for day-to-day, responsive building repairs) needs to be tackled in a different manner. In such a contract, works are called upon as they are needed so the list of works or tasks in the specification has to be as broad as possible so that *all potential repairs* are covered. It is usual to specify these works as complete tasks to be priced on a Schedule of Rates (see later) and to be as clear and specific within the definition of each individual rate as is possible. Remember – what you ask for is what you will get. If you do not specify to re-connect the taps when they have replaced a wash-hand basin, they will not be reconnected!

Handy tip: In such contracts (and this tip is handy for most contracts), always include a schedule that covers an hourly or daily rate for staff or operatives so that, should any task be found that lies outside of the specification, the tendered rates can be applied and thereby costs controlled.

Specification Writing – Meeting the Need

The need is met through the Specification and there are various forms a contract specification can take. The one you use will depend on the nature of the contract or project. In general, specifications can be Input or Output based:

- *Input specifications* tell you what has to be done to achieve the end result (e.g. 'lay so may bricks in a certain manner to build a boundary wall – if you do it this way, the product will meet our needs').

- *Output specifications* tell you what is wanted at the end of the job (the 'product') and leave the detail of the how it is to be done to the bidder. These are not often used in construction but are common in service or maintenance contracts (e.g. 'the grass is to be kept to a length not exceeding 50mm at any time – how you achieve this is up to you, the contractor').

Input Specifications

Where the process tends to be critical to the quality of the outcome, an Input specification is often the preferred option. It tends to create an easier contract to manage on the basis that the process is clearly defined and so any deviation from the requirement is more easily identified. An example of this might be the renewal of a kitchen wall cupboard.

In some detail, the specification might identify removal of the old unit; assembly and re-installation of the new positioned to match the existing units, ensuring doors are fitted and properly operational, making good any damage to walls and décor and leaving the area clean and tidy upon completion.

It picks up on key areas of the process that have to be done to ensure a good job at the end of it. If any of the constituent parts of the specification were omitted, the job would not be good, even though it might have met the specification as written. The moral here is that this type of specification needs to be thorough in content: note in the above example I did not specify fitting any door furniture – does that leave room for the operative to leave site without putting a handle on the door?[7]

---

7. Reasonableness would deem that the door would not be 'operational' without a handle, but the specification ought to include it to ensure a 'good' specification and save disputes.

Output Specification

Output specifications tell you what the client wants to see at the end of the job, whilst the details of the 'how' are of minimal interest. An example of this might be: "Tennis courts to be kept free of litter and swept free of dirt, ready for use at all times".

The client does not mind if you sweep the litter, pick it up, bag it, use a machine, use a broom or whatever, just so long as it is all kept clean and ready for use. The end result is sufficient to determine that the outcome is as the client (and the customers) require.

Specification Format

The format of the specification is key to the way the tender will be submitted and the contract itself delivered and managed. It is therefore essential that we look at specification formats in some detail.

Input and output specifications can each themselves be formatted differently, depending on the requirements of the situation. The format may be determined by the complexity of your requirements, ease of evaluating the tender submissions or perhaps the degree of flexibility required in the way the contract is delivered and managed once on site.

In time, the best format to use for a particular type of contract will be obvious to you, but in the meantime you may need to ask people with the appropriate experience, such as your QS.

The three main formats comprise:

- Written Specification.
- Schedule of Rates (SOR).
- Bill of Quantities (BOQ).

We shall look at these in some detail, taking a simple 'construction' project and specifying it in each of the three ways. All forms of specification writing can, in general, be applied to any job but some are very much more suitable than others for particular situations.

We will indicate the suitability and limitations of the different forms as we go.

The sample project is to completely renovate all the rooms in a sheltered housing unit (SHU), assumed to be unoccupied at the time of the work: renovation work is to include all tasks, from hacking off plaster where necessary to applying the last coats of paint and tidying up prior to leaving site.

For obvious reasons, I will not draw up a *complete* specification each time, but a chunk sufficient to make the format and usage clear.

1. Written Specification

Normally used for small works projects, this format 'talks' the bidder through the whole process of the project or job. It may literally describe the job from beginning to end or it may divide the description of works up into chunks by trade (for example) or by room or area.

Pricing work that has been specified in this manner can be done either as one lump sum or per part, depending on the pricing document and requirements laid down in the instructions to tender. The client may provide the quantities or measurements for which the price should be submitted, or the bidder/s may have to study drawings and visit the site in order to assess the size of the job for themselves. If the client specifies quantities, any errors will be down to them and the tendered sum will probably need to be re-negotiated for any element(s) where there is an error. If the contractor is to specify quantities, all bidders have to have access to the site for measuring. You can give measurements and say you accept no responsibility for them – the bidder then has to check them. Your choice. How do you want to do it?

A written specification has to be specific *(sic)*, as it has to describe the process in sufficient detail to enable the bidder to offer a (fixed) price for all required works to complete the job. Any changes to the requirements will necessitate negotiation to agree a cost variation (may be up or down). Any unknowns in the specification will necessitate the use of provisional sums which, by definition, may or may not be accurate: actual costs will have to be agreed on site once the quantities are known (may be labour and/or materials costs).

In our example above, as there are many rooms, it is anticipated that each room would require an individual specification as not all rooms would be in the same (starting) condition and so would require different amounts of work. The returning bids would reflect this.

Example:
"Bedroom #1:
Strip all wallpaper.
Hack off 4m² plaster (as marked) to brickwork, bond and set to finish flush with existing.
Thoroughly prepare all woodwork (including windows, surrounds, skirting board, picture rails and architrave) – burn off all loose and flaking paint (provisional sum) and sand all surfaces ready for painting. Apply primer to all bare surfaces, undercoat overall and apply one white gloss topcoat to finish.
All walls to be sealed and horizontal-lined with 60gsm lining paper; apply one mis-coat and two topcoats of satin Magnolia emulsion to finish.
Ceiling to be washed and filled as necessary, sanded smooth and finished with two full-bodied coats of brilliant white matt emulsion. Etc..."

This written specification could include or be followed by:
- Specification of any of the materials where required (e.g. "Dulux Brilliant white Gloss BS number: XXXX or equivalent[8]"). If not specified, the contractor will be able to

---

8. Note use of "or equivalent" – the specific identification of a named product can be challenged as uncompetitive, so you must always allow bidders to put forward alternative brands or models, but you can insist they prove the product's ability to meet the performance you have specified (in this case, is the paint they propose using as 'good' as the Dulux product?).

choose the products, and will normally choose the cheapest to keep the tender costs down and the profits up.

- List of provisional sums ("For burning off woodwork to bare timber, PS £270, labour and materials")
- Inclusion of the obvious, such as "…all access equipment to be supplied and included in the price; … all waste and materials to be removed from site upon completion", etc.

The price submitted would be for all the work items listed above, including materials, through to completion of the job.

Note: The above is an 'input' or 'conformance' specification, listing the steps or activities and materials needed to arrive at the required conclusion. An exaggerated example of an Output Specification for this project could be: "Carry out all renovation and decorative works necessary to leave the room clean, fresh and ready to let".

The obvious problems with an output specification for this job could be:

- No definition of 'ready to let'.
- No specification for standard of finish.
- Underlying quality of workmanship is not specified.
- No specification for style or colours of finish.
- No specification of materials.
- Comparison of different quotes is difficult.
- Quality of the 'finish' is not specified, so
- The standard of the finish is not guaranteed, leading to
- Disputes regarding the final job.

An output specification is therefore not ideally suited to this type of project.

2. Schedule of Rates (SOR)

Usually used for larger projects and Term Contracts, a Schedule of Rates is a pricing document that lists individual tasks and asks for prices against each one. This enables prices to be submitted for the anticipated tasks in such a way that they can be called upon (or omitted) as required. The main advantage is therefore that precise quantities need not be known before the instruction to proceed is given and changes to the specification or unexpected problems can be catered-for within the specification and quoted sums.

It is therefore best-suited to call-off contracts (such as maintenance and repair services) where tasks may be required on demand, or (less ideally) works contracts where quantities are not certain at time of tender. For this reason, this method of specification also lends itself nicely to tenders incorporating early contractor involvement.

Additionally, the client can leave the price elements blank for the bidder to complete, or can enter prices for each item and ask bidders to submit an uplift or discount figure across the board to arrive at the bid value. In my examples below, I have assumed the contractor enters the price – which is the most common approach in works project tenders. For term maintenance and repair contracts, it is more common to enter the prices and seek tendered discounts or uplifts. There is much more on this in Chapter 7.

However, behind each individual SOR there has to be a very detailed item specification describing what each one means, including the labour element as well as the parts and materials. By its nature, a Schedule of Rates specification (or tender document[9]) can be very bulky and time-consuming to both prepare and submit as a bid.

If you are creating it from scratch, it will also be expensive.

It is worth noting that a SOR specification (and a Bill of Quantities – see later) will require a preceding element of descriptive text or narrative, to set the SORs into context: basically, builders will need to know what they are building!

There are various conventions and points to note about the format of a SOR Specification:

---

9. A 'Schedule of Rates' is also a format in which tendered prices are submitted.

- Certain areas of work are normally priced per square metre (m²), in particular:
  - Wall work (e.g. papering, tiling, plastering, brickwork)
  - Horizontal surfaces (e.g. paving, turf, asphalt, felt, flooring)
  - Windows (supply, fit, paint, overhaul, etc)
- Some areas of work are normally priced per linear metre (lm):
  - Joinery (e.g. skirting, handrail, picture rail)
  - Pipework
  - Guttering
- Some areas of work may be priced by volume (m³):
  - Concrete pour work
  - Land fill
  - Top soil
  - Removal of demolition arisings and other waste
- Division *can* be into any units you choose so long as they are defined (see later).
- SORs can be as comprehensive or 'single-minded' as you choose.
- A SOR item can cover just one simple task or a range of tasks.
- Where a 'complex' SOR results in a completed minor job (e.g. the installation of a complete replacement sink unit), they are often referred to as Standard Order Descriptions (SODs), but are compiled and priced in the same manner as a SOR.
- SORs can detail the specification or there can be a separate section clearly defining the required quality of work and materials for each task.
- SORs can be labour only but most commonly include supply of the necessary materials.
- It is normal to offer different rates for different volumes of the same task. Normally, three volume bands (or columns) are set up, each covering a range of quantities, so as to allow

the submission of prices reflecting any volume discount. The actual volume break-down should logically reflect the scope and needs of the job.

Two examples of a Schedule of Rates follow. The first *(Example A)* shows the most common format, breaking the work down into individual tasks that can be applied to each of our SHU's rooms in turn.

The second *(Example B)* shows how an overall specification could be set out for all the rooms – similar in format to the written specification. The tendering criterion would then become the price per room, which would in turn depend upon the size of the room in question. The size of the room could then be measured by floor area (most common method) or by volume (a better method if the building has particularly high ceilings, for example).

Whichever method is chosen, it has to be fair to all parties – bidder and client alike.

[In the following examples, bidders would enter their tender price in the boxes marked 'P'].

Example A: Individual tasks

Description		1 to 5 m²	5.1 to 20 m²	Over 20 m²
Wet Trades				
Hack off plaster up to 20mm thick, bag up arisings and dispose from site.		P	P	P
Hack off plaster exceeding 20mm, rate per additional 6mm thickness.		P	P	Etc.
Apply bonding to bare brickwork, up to 20mm thick.				
Apply skim coat of multi-finish plaster up to 3mm thick.				
Etc.				
Decorating – walls and ceilings.				
Strip wallpaper using steam or water, bag arisings and dispose off site.				
Seal wall using one coat of Size as per manufacturer's instructions.				
Line walls horizontally using minimum 2000 Grade lining paper and paste according to manufacturer's instructions.				
Etc.				

Decorating – previously-painted woodwork.	1 to 5 lm*	5.1 to 20 lm	Over 20 lm
Prepare timber by burning off, sanding, washing with sugar soap and rinsing, filling and sanding as necessary ready to receive paint.			
Apply primer to bare wood, and one undercoat and two white gloss top coats overall.			
Etc.			
Decorating – previously-painted wooden windows.	1 to 1.5 m²	1.6 to 2.0 m²	Over 2.0 m²
Prepare timber by burning off, sanding, washing with sugar soap, rinsing, filling and sanding as necessary, ready to receive paint.			
Apply primer to bare wood, and one undercoat and two white gloss top coats overall.			
Etc.			

*lm – linear metre

Figure 6.3 – Example of a Schedule of Rates

Example B: Whole-room schedule

Renovation – whole room (floor area)	Up to 6 m²*	6 to 12 m²	Over 12 m²
To renovate any room in line with specification, including all preparations and finishes, leaving clean and ready to furnish, with all waste removed from site.	P	P	P
Etc.			

*Floor areas but could be m³ for room volumes

Figure 6.4 – Schedule of Rates using Whole Room Units

3. Bill of Quantities (BOQ)

A BOQ is similar to a SOR specification, but the anticipated quantities for the whole project are set out for pricing by the bidder. Again, a clear specification of all items is required, normally located with but not within the BOQ descriptions.

This format has the advantage of establishing a more definite price for a job, but is obviously only suitable for project work – it could not be used for term or call-off contracts. BOQs take longer to compile as the Surveyor and the QS have to work out the quantities for the whole job before it even goes out to tender. They do not as readily allow for additions and deductions as a SOR-based contract, but the breakdown of costs within the Bill do make quantity changes easier to cost than a written specification.

Nevertheless, BOQ-priced jobs will tend to work out cheaper because the bidders, who submit a price against each quantified item, are more assured of the quantities involved than they are with a project priced on SORs.

In our SHU example, we would add up the required quantities for all the rooms in the SHU (i.e. the whole job). This would be the best method in practice, as it would attract the best price. However, it could be set out on an individual room by room or floor by floor basis. Such a format would leave more scope (for example) for taking individual rooms out of the job if the price exceeded budget and it would also make the tendering process more accessible to smaller providers if the floors were each tendered as separate Lots (see Chapter Three). Such considerations would form part of your tendering strategy.

In a BOQ, 'Qty' is the total quantity anticipated in that particular 'Bill' or job (i.e. the whole project) and 'Rate' is the bidder's rate per unit measure – not always included. 'Price' is the actual bidder's tender for that work item for the whole project as listed.

Example (overleaf):

Description	Qty	Unit	Rate	Price
Wet Trades – walls				
Hack of plaster up to 20mm thick, bag up arisings and dispose from site	125	m²	P	P
Hack of plaster up to 30mm thick, bag up arisings and dispose from site	32	m²	P	P
Apply minimum 20mm bonding coat to bare brickwork, ready for finish, including all beads	157	m²		
Apply skim coat of multi-finish up to 3mm thick	157	m²		
Etc.				
Decorating – walls and ceilings				
Strip wallpaper using steam or water, bag arisings and dispose off site.	865	m²		
Etc.				
Decorating – previously-painted woodwork				
Prepare timber, by burning off, sanding, washing with sugar soap and rinsing, filling and sanding as necessary ready for paint	480	lm		
Etc.				

Figure 6.5 – Example of a Bill of Quantities

Definitions

Whatever form of specification is used, a list of definitions will be required and the complexity of these will depend on the needs of the contract. Definitions explain what you mean by the words and phrases you have used. In the example I used of cleaning a tennis court under Output Specification above, "dirt" may have to be defined so as to distinguish it from "dust" (yes, you cannot expect an outside tennis court to be kept free of 'dust' and the contract, through the specification, must make this clear and your demands reasonable).

Never assume that the meaning of a term is obvious and common to everyone who reads it – arguments will certainly ensue. Make sure every term you apply that is key to service delivery is clearly defined within your documentation, although there will have to be a balance struck between adequate definitions and pedantry, guided by experience and the criticality of the item at stake.

Standards of Product

Rather than bog the specification down with repeated phrases, the standard of the products to be used on your contract – the quality aspect – can often be generically defined.

You must remember that the higher the specification (i.e. the more you ask for or the higher the standard you demand) the more you will pay. For most Council contracts, the need is for work and materials of 'a sound standard'. To this end, materials should be specified to meet known quality measures (such as BS or ISO), where the quality of the product will be assured and verifiable on site if needs be.

You must never specify a particular brand or product directly and explicitly. You can specify a brand or product as an example of your needs, provided you include the phrase "or equivalent." The bidder may then furnish you with an alternative product and, provided it meets the same performance specification, you are obliged to accept it as part of their tender.

The only exception to this 'rule' will be technical specifications that demand certain performances (such as from a supporting beam, for example) or that identify a product where a deal for its supply has already been agreed elsewhere or where the product has to match some existing installation or infrastructure. For example, the Council may have negotiated a deal for the supply of a specific boiler and wishes to standardise on this model throughout its properties.

In such a case, there has to be a sound and justifiable reason for the explicitness of the specification and in many instances there will be a mechanism whereby the client purchases the product and the tendered work is for installation and consumables only.[10]

Specifying labour or process standards

Just as the standard of materials can be defined so, too, must the standard of workmanship or a process be specified, and this is

10. If a client-specified unit is to be supplied by the contractor, it is normal for a management or handling charge to be added at time of tender to any client-negotiated price. This may be just 5% or up to 25% depending on the nature of the product, delivery logistics and any allocation of warranty responsibility (i.e. risk) laid off onto the contractor.

most easily done where industry standards prevail. As an example, shrub-pruning methods and timings are clearly laid down by the Royal Horticultural Society and these can either be quoted or simply referred to as the standard of workmanship to be met at all times.

You will find that many trades – like horticulture above – have clear guidelines on good practice whilst some (usually the ones you are working on, of course) do not. For this reason, you may well find that you will have to define the quality of workmanship you are expecting and make it clear in your specification.

Whatever type of specification is being used, you will need to ensure that your requirements are clear, complete and comprehensive and you should include drawings or other aids where appropriate. Always work on the principle that you will only get what you ask for – obviously, whatever you leave out will not be provided.

Accreditations

For certain trades or categories of work, specific accreditations may be legally required (e.g. GAS SAFE and NICEIC) or desirable (e.g. FENSA)[11]. Where they are not statutory requirements, the value of providers having such accreditations is for you to judge.

The Impact of Time

It must be remembered that the time you specify for a job to be completed in will impact on the cost – and possibly the quality. If you specify too short-a time, the contractor will need to allow for overtime or shift allowances or maybe additional lighting. If they have to push to meet an over-tight deadline, the quality of the product may also suffer.

If you allow too long for the project, Preliminary[12] costs will be higher.

There are certain key factors that need to be taken into account

---

11. GAS SAFE – Work on gas supplies; NICEIC – Electrical work; FENSA – windows
12. Preliminaries – costs of being on site such as staff facilities, scaffolding, security, etc.

when considering project-times: What is a *reasonable* period of time to complete the work? Will the weather impact on the time? Do you have tenants waiting to move back in? Is the work linked to a school holiday? Does the contract time cost you money? Are you going to impose LADs[13] for over-runs against scheduled timescales? Due consideration needs to be given to the time specified for the project within the tender documentation so that your requirements are reasonable and balanced.

With a term repairs contract, time comprises the speed with which the provider has to attend to carry out the repair (the response time) and / or the time in which it has to be completed. The more stringent either of these are, the more each repair will cost. For this reason, such a contract has a range of timescales, with different rates for each normally established through a percentage premium or uplift for the faster times. In this way, you will pay more for the urgent jobs, but will not have to pay so much for jobs that are less urgent or need a period of time for materials to be acquired (for example, a replacement window has to be measured and made).

Where time is not so much something you wish to specify, but more a matter of what a contractor can offer, a proposed project programme will form part of the quality aspect of a MEAT tender evaluation model as discussed earlier.

Specification Writing – Added Value

This is the hardest element on which to provide concise guidance. What I say will hopefully stimulate thought and indicate the areas in which added value may be achieved, but further and broader investigation of your own is recommended. Always bear in mind, however, that 'added value' can be dangerous territory. When we come to the topic of evaluation, you will see that setting added value as a requirement of a contract and then as something to

13. LAD – Liquidated and Ascertained Damages constitute claims that offset *actual and cashable* losses against sums paid to providers at time of Certification. They cannot be levied as penalties. See Chapter 15.

be considered at tender evaluation time is very sticky ground indeed.

What is 'Added Value'?

Added value can be defined as what a contractor offers within their bid (or during the course of the contract) that will add to the quality and effectiveness of the finished product but is outside of the actual specification. Normally, added value contributes to the product or the community on a local or broader scale and can take many forms, but it will generally reflect an *investment* in the area in which the contractors are working. What should be sought is something *back* from the contractor over and above simply the finished, specified product.

The Social Value Act requires you to consider such (added value) gains for Services Contracts but – oddly – not with works contracts. This is odd because often works contracts over many opportunities for such gains. Often, sustainability measures come in as added value and Chapter 13 looks at this in greater detail, but we will give it some consideration here.

To properly assess Added Value within a product, Whole Life Costs (WLC) need to be taken into account and these should include the cost or value of a product throughout and at the end of its useful life. WLC is covered in some detail in Chapter 13.

What contracts can have added value?

It is obvious that smaller, one-off projects have little leeway within their cost and time constraints and provide little opportunity for investment by the contractor in anything outside of the work in hand. Thus, Added Value is more easily achieved in larger projects that have a longer commitment or a wider scope. A perfect example of this is a regeneration project, which will be broader in perspective, take longer to complete and will tend to embed the contractor into the immediate community.

Similarly, Partnering contracts and Term contracts are ideal vehicles through which to seek Added Value, and the possibility should be considered as part of the project plan for any procurement exercise.

Examples of added value

A simple example of added value would be the establishment of a community fund – or even the construction of a community hall – as part of a Neighbourhood or estate redevelopment scheme. Ostensibly donated by the contractor, it is obvious that they will afford the 'gift' out of the proceeds of their bid. However, in terms of Added Value, the donation will not have been specified within the tender documents and is offered through the contractor's good will as an item that adds value and quality over and above the finished product and for the benefit of the community.

Added value 'items' can take many forms and a key area in which added value can and should be sought is *Sustainability*.

Sustainability – covered in Chapter 13 – has many forms but essentially looks to the longer-term, altruistic benefits of the community and society at large. Sustainability goes beyond issues surrounding climate change and the use of local labour is an example of how a contract can help both the local economy and social demographics. Taking this a step further, upskilling local labour by establishing a local labour training scheme or centre would be even more effective and actually establish a lasting – i.e. sustainable – legacy for the future of the community, effective beyond the end of the project.

Ecologically, contractors could undertake to provide (or aim to provide) all power to their site offices using solar panels, to reduce waste and re-schedule deliveries to minimise vehicle miles. Social integration and sustainability could be engendered through the organisation of community events linked to consultation exercises and the involvement of schools in competitions to decorate hoardings, design play areas, etc.

The price of added value

Of course, none of this comes free – but contractors will be willing to offer added value over and above the tendered sum if they see that the investment is worth their while. It will be worthwhile if it generates good will with the client and the stakeholders, makes the job run smoothly, provides good PR and stands them in good stead for future opportunities.

Also, it is a reality that many if not most contractors (I would hesitate to say "all") do have a social conscience and respond to what is now being called *Responsible Procurement*. Responsible Procurement (as in Corporate Social Responsibility) is the umbrella under which such issues as climate change and social and economic development are considered to be matters that can and need to be addressed within the procurement process.

Remember – MEAT criteria allow you to seek added value and not automatically appoint the lowest bidder. You do not always need to specify what added value items you expect out of a contract: always leave bidders scope to bring new ideas to the table. Some of them are very good at it.

One point – and this is the (very) thin ice in the procurement rink – beware the evaluation! It is always possible to include benefits such as those we have suggested in the tender evaluation model, but justifying why one bidder's proposal is worth more than another's because of their added value could lead you into trouble. If an unsuccessful bidder challenges the outcome of your evaluation, you could literally have to stand up in court and justify your decision. Justifying a decision based on intangibles may be difficult.

Models that set values on such additional benefits are becoming more common – and accepted – but my recommendation for now is to encourage added value at time of tender, avoid using it as an evaluation criterion and then pursue it once the contract is awarded and on site. Meanwhile, see the chapters on Challenges and Evaluations.

Contract Manageability

Do not fool yourself – no matter how good your procurement, if the management of the contract is rubbish, so will the outcome be.

An effective and successful contract is one that is managed well and the key to effective contract management is ensuring that the correct management tools are built into the contract in the first place. Whilst the Terms and Conditions lay down the over-arching rules of the contract, it is the specification that defines what work needs to be done – i.e. the product, how it is to be delivered and what the client can do on a day-to-day basis to ensure it is achieved.

I refer you back to our circle showing the procurement cycle and remind you of how contract management is a key part of this cycle, feeding into the procurement process and ensuring its needs are met within it.

For effective contract management, the specification has to be clear, specific and reasonable (i.e. deliverable) and it needs to be – so far as is possible – beyond contention whether the required output has been achieved. Avoid jargon so far as is possible and check for any contradictions.

Risk Management

Risk management is covered in Chapter 12, but its place needs to be considered here. The process of risk management will consider threats not only to the procurement process but also to the smooth operation of the contract itself, once let.

These threats to the contract may be matters that, once identified, can be mitigated at the procurement stage. For example, the risk of a provider going out of business could be mitigated by the establishment of a Performance Bond (see Chapter 16) and/or a back-up provider and the problem is then – so far as is possible – resolved.

Some problems, however, cannot be so neatly resolved and the procurement process has, instead, to put in place the tools to deal with the potential problems as and when they arise. An example of this could be the provider failing to perform to the required standard. Procurement in this instance will have to ensure that measures are laid down within the contract to deal with this.

Thus the Risk Management process will assist with the writing of the contract specification and the terms and conditions under which the contract will operate, and should be used as a tool for this purpose.

Key Performance Indicators (KPIs)
It is sometimes necessary to have specifics against which delivery can be measured – Performance Indicators (PIs). There are sets of standard, national indicators used within the construction industry[14] as well as Best Value KPIs (Key Performance Indicators). Benchmarking Clubs enable similar client bodies (e.g. Local Authorities) to share common PIs and so compare their contractors' performances (and thereby their contract management standards).

The 'published' PIs may or may not provide the information you require to monitor your contract, or monitor things in the manner you want, so you may have to construct your own PIs to meet your specific needs. These are often called *local* PIs.

You need to include PIs clearly within the specification and make it known when and how the data is to be collected and then evaluated. As a tool, remember that PIs are not an awful lot of use if there is no outcome from missing or meeting the targets – so some form of (legal!) retribution or reward is required.

For long-running or ongoing contracts (i.e. not usually possible in construction project contracts) it is advisable to set the targets marginally higher for each succeeding year so as to encourage continuous improvement as advocated within Best Value.

A detailed discussion on PIs is not an area that strictly falls within the scope of this book. However, be advised: it is not simply a matter of thinking of a good thing you want a provider to achieve and writing it down! PIs need to be relevant and required and should meet SMART[15] criteria; you should be able to measure, collect and evaluate them and then use them to better-manage the

14. See the Constructing Excellence website
15. *SMART* Targets are Specific, Measurable, Agreed, Realistic and Time-related. In the context of procurement, discussion to 'Agree' the PIs with the contractor is not really an option, but the over-arching principle of reasonableness applies.

contract. As a procuring officer, your main job is to see that the client's PIs are included properly in the contract documentation, but you will need to know enough about the subject to see that the proposals make sense. If you *do* need to draw them up (you may be the client as well!) and are not sure, be advised and seek help from a QS and others who have had previous experience of drawing up and managing PIs. You will find more on KPIs and their use in Chapter 17 on Contract Management.

Other tools

Other tools for effective contract management that need to be covered in the specification include:

- Clear management processes, including
- Regular contract meetings
- Default processes for failure to deliver, with procedures for remedies that include
- Charging administration costs for dealing with or processing under performance. These cannot be penalties.
- Levying of LADs[16] to offset accrued losses brought about by under performance, e.g. loss of rent income on a property.

Action in times of failure

You need to be aware that it is not legal to include penalties in your contract, so any failure on the part of the provider needs to be addressed in a manner that is reasonable and not considered punitive.

In term contracts, a process involving written Default Notices is common. Default notices identify the failure and seek (demand) a remedy within a given period (maybe hours or days depending on the nature of the failure and the needs of the contract). The process has to be progressive: starting at, say, 'get it sorted within one working day' and culminating in the appointment of another provider (the appointed back-up, if there is one) to resolve the issue. There is normally a middle stage that gives the contractor

---

16. LADs – see earlier: Liquidated and Ascertained Damages

a written warning of a third party being employed to resolve the failure.

Provided it is written into the contract, you are entitled to recover the costs associated with employing another provider (including your 'administration' costs of doing this) and you could also apply administration charges to the process of simply raising a Default Notice. Some contracts apply a points system to Defaults and seek an overall financial settlement at regular intervals (quarterly, for example) against the contract sum on the basis that the service has not been provided.

In all cases, the process must be made clear in the contract documentation and it is always wise to have a view from the Legal member of your Team on your proposals.

Always remember *reasonableness*. Reasonableness has to apply in several directions and contracts are managed well or badly depending on how reasonable all parties are with each other. It is good practice to avoid issuing Defaults unless they are really necessary; Default remedy periods should be a balance between the needs of the service and an appropriate period of time to correct failures; any administration or other charges should be real and accountable as they fall into the category of LADs – they are *not* fines.

Finally, it is equally reasonable to have rewards in a contract as well as punishments. It is possible to include cash bonus systems (tricky to provide a business case for this, especially in such austere times, but it is possible) and offsets against Default points, for example.

In terms of procurement, the rule is to ensure that means of restitution are meaningful, usable and clearly defined within the contract documentation.

On the Ground
You need to ensure – as much as you can – that what you put into the specification is translated into practice once the contract starts. Referring again to the 'cycle' diagram at the beginning of this chapter, you will see that a key stage in this translation

process is training. Particularly with term contract management, training the staff who are to manage the contract is a key factor in ensuring that the care and thought you put into it's execution are not wasted.

The training need only be for an hour, but should cover all the salient points regarding what the contract is designed to provide, how it is to be provided and what managers can do if the provider fails to meet these requirements. It should cover ordering processes, remedial actions, and any major changes to any previous contract. This training should be accompanied by a comprehensive Contract Guide or Manual, written by yourself, that gives *all* the necessary details of the new provision, including contractor names and contact details, Lots and Packages, contract periods, how to manage it, key clauses from the specification and terms and conditions, and so on. It really *should be* a manual on how to run the contract and should almost be sufficient in its own right.

It is also beneficial to offer similar training to the providers so that their supervisory staff know what to expect and have the opportunity to ask questions.

With project contracts, obviously, a Manual is not required, and often the person managing the contract will have been involved in writing it so such training is not generally required.

All of this should be supported by *ownership*: do not 'build the boat' then push it off from the quayside, never to see it again. Keep a hold of it throughout its life and learn from its operation; be on hand to provide help and assistance when any management queries arise – for arise they will. Make sure your name and number are in any Guidance Manual you create and accept all feedback and criticism as constructive and use it to improve next time.

Summary
- The Specification is the tool that will get you what you want from a contract.
- A specification can be written in many formats.

- The specification must include contract management tools.
- A specification takes on board budget, value for money, the need, meeting the need and any added value.
- Use others' experience to draw up a specification.
- Have and imbue ownership.

7 // Evaluation Models and Evaluation Techniques

A Most Important Part

In Chapter Four I made it quite clear how important the timetable is to a successful procurement: it is absolutely vital. So, too, is your evaluation model, be it for a Questionnaire or the actual tender.

The evaluation model is *the thing* that will ensure you get the right provider on board. If the contract is not great and the specification is not the best, if you have the right contractor on board it will almost certainly still work. You can have the *best* contract and a perfect specification, but if you do not have the right contractor on board it will be a mess and a nightmare to run.

On this basis, it is hard to put too much effort into an evaluation model. Even if it is only a couple of A4 pages long, it could well be *the* most important document in the whole tender pack. Remember that.

As I have said, 'evaluation' comes at tender stage *and* at the Questionnaire stage and we shall look at both. An obvious statement is that price is not assessed at the Questionnaire stage. However, the Questionnaire stage *does* involve assessing responses to questions pertaining to quality aspects of a firm[1], so the advice on assessing quality submissions *will* apply to both. Where any differences do occur, I will point them out.

The regulations are quite clear that for tenders with a value below the Services threshold (remember - even for works contracts), the open procedure must be used, i.e. you cannot pre-select. However, you can still undertake assessing a bidder's suitability and capability to carry out the work and you do this – yes – through a Questionnaire, but this time issued *with* the tender documents and based on a pass/fail basis. The stipulation is that questions *must* relate to your contract, but then we would expect it to, anyway.

---

1. Except for tenders below the EU Services threshold – remember?

They do, in fact, clarify this a little by stating that the questions you ask must not be more onerous than necessary (although the TFEU requirement for proportionality already covers this) and add that you cannot require, in any tender, that a bidder has an annual turnover of more than twice the value of the contract[2] (and this applies to tenders of any value, above or below threshold). The Cabinet Office stipulates the use of Standard PQQ questions where they are applicable. This does not preclude the use of your own questions so long as they do not contradict the standard ones and are overtly relevant to the contract being tendered.

You are referred to Cabinet Guidance at: https://www.gov.uk/government/publications/public-contracts-regulations-2015-requirements-on-pre-qualification-questionnaires

Types of Tender Evaluation

Within the EU Regulations, there is only one way to evaluate bids: Most Economically Advantageous Tender (MEAT). The EU Regulations *used* to allow lowest price as well but that has now been changed – sort of. The EU now recognises that, depending on the circumstances, you could in fact consider the lowest price to represent the most economically advantageous tender and... bingo! it's back in again. So, as before, we have the two types of evaluation principles recognised under the EU regime. I shall continue to refer to 'lowest price' to distinguish those evaluations where quality does not feature in the scoring mechanism.

Let us now look at these two types.

Lowest Price

Lowest Price assesses just that – "Which is the cheapest offer?" As nothing other than price is considered, *everything* else about the requirement (such as speed, frequency, size, colour, material, shape, and so on) all have to be specified within the tender documents with no room for ambiguity.

---

2. Use the annual value of a term contract or the equivalent sum of all Lots for which they might be bidding.

For this reason, lowest-price tenders can only be run where the specification and product (or quality) management can be specific. The purchase of items (such as printers) can easily be tendered in this way as all aspects of the purchase requirements can be clearly stated.

Oddly, and against natural intuition, many building projects – particularly smaller ones – can also be tendered using lowest-price submissions because the combination of a specification, strict Building Regulations and numerous controls on site can all work together to ensure the quality of the final product.

The evaluation model is therefore quite simple – list the bids in order and accept the lowest. In practice, of course, it is not quite as simple as that (this is, after all, procurement). Except for the simplest of tenders, you will probably need to have a QS on board to assist with the evaluation. This is particularly so when the submission consists of several, individually priced elements that go to make up a bottom-line total. A QS should check that all elements have been priced 'sensibly', they should arithmetically check all bids and confirm that the lowest bid is a viable offer and finally confirm that the bottom-line figure *is* the total of all the elements.

If the lowest bid is very much lower than all the others, and especially if the other bids are fairly well grouped together within a range, you do have reason to suspect that the 'winning' bid is lower than it ought to be – a 'rogue' bid, known as an abnormally low tender (ALT). The Regulations now require you to investigate and confirm all abnormally low bids rather than simply reject them or – as some might have done in the past – simply accept them.

Rogue bids can happen for several reasons:

- Arithmetical errors on the part of the bidder (perhaps missing something out or an error in a spreadsheet formula)
- Decimal points or commas in the wrong place (typographical errors do occur)
- Omissions
- Deliberately low price (known as a 'suicide bid' or 'buying' the contract) due to:

- Shortage of work
- Need for cashflow, almost at any price.
- They are actually able to do the job more cheaply than you expected.

Errors

When there is an arithmetical error (including a misplaced decimal point) the mistake will be identified and what is known as 'Alternative 1 of the NJCC's[3] Code of Procedure for Single Stage Selective Tendering 1996' can be invoked. This approach is not compulsory, but is fairly common practice – especially in construction, where it evolved.

When using this option, numerical mistakes can make a bid either erroneously high or low and the lower amount always applies, be it the original submission or the corrected version. When the error is discovered, you should contact the bidder, point out the error and confirm the corrected final tender sum with them. The bidder then has the choice of either accepting the lower figure or withdrawing their bid. Sometimes this can seem a bit harsh, but this is one of the standard rules of the game, if you choose to adopt it.

Instead, you may allow the correction of errors, albeit without going so far as to re-write the bid. The key point about allowing corrections is to make sure you are being equally accommodating to all bidders in your degree of 'tolerance'.

You must ensure your Instructions to Tenderers clearly state what process you will employ as part of the evaluation process to deal with errors. If you are using Alternative 1, you need to make it absolutely clear in your evaluation information document and advise of its implications.

Omissions

Omissions – leaving bits out by mistake or otherwise – will make a bid seem irrationally low. You have three choices here: either to

---

3. National Joint Consultative Committee of Architects, Quantity Surveyors and Builders

invoke Alternative 1 (on the grounds that they have added up all the bits wrongly by missing some numbers out) or to declare the bid non-compliant on the grounds that they haven't bid against what they were asked – i.e. haven't done what they were told or you can allow a correction of the bid. Again – provided you have advised in the tender documents what your approach will be, they are all acceptable options.

Deliberately low price

If the figures are all present and correct but the bid is considered to be exceptionally low, you must contact the bidder, confirm the tendered sum with them and ask if they wish to stand by the sum tendered. The bidder then has two choices: stand by the sum they tendered or withdraw the bid.

If the bidder stands by an abnormally low tender sum, *you* then have two choices: accept it or reject it. There have been various court cases regarding the acceptance or rejection of low bids, so this is not an unusual situation. You can reject a bid if you have reasonable grounds to think that the bidder cannot provide the service for the sum tendered. This is not always an easy call: you may have to defend your decision in a court of law, so certainty is required. The QS should be able to build up a case based on known market labour and materials costs, and may be able to provide evidence of low or zero profits, and this may well suffice.

Before you finally reject a tender, it is a sensible strategy to point out your concerns to the bidder and ask them to demonstrate how (or why) they can fulfil their obligations under the contract for the sum quoted. This may call their bluff (if it is one), enable them to demonstrate that they *can* do the work for the money or provide enough evidence for you to reject the bid with confidence.

But having said this, do take care. I encountered an instance where a company 'bought' a contract and we were minded to reject their exceptionally low bid. However, when asked, the bidder explained that they wanted a foothold to enable them to expand into the region. We accepted the explanation, the contract was completed very successfully and we would have been wrong to reject the bid. There is no 'rule' that fits all – it is a judgement call.

'Buying' a contract may be a reasonable strategy to overcome a (perceived as temporary) shortage of work. The aim would be to cover overheads but keep the tender sum low by minimising or even eliminating the profit margin. When it is done simply to generate cashflow, this is often a sign of a firm in financial difficulties and should be taken as a warning sign to steer clear.

Firms will not often tell you if they are in financial difficulties so where you have a suspicion, you should undertake some financial checks through one of the known credit-rating agencies and you can even ask the bidder for a set of their latest (emphasise 'latest') accounts. If, after this, you are still mindful to award the contract to them you can put in place additional surety through the provision of a Performance Bond or a Parent Company Guarantee to cover your organisation's costs if they do go 'under'. Such assurances may not completely cover any ensuing losses but they should help to mitigate them.

Bonds and Guarantees

Bonds and Guarantees are covered in detail in Chapter 16 but we shall look at them in the context of tender sums here as well. Performance Bonds are provided by the appointed contractor to cover your costs in the event of their being unable to complete the contract. Such failure is most often due to the company going into liquidation, but other causes can sometimes be to blame. Bonds can be supplied by the company's insurers or their bank or they can be obtained (i.e. bought) from professional trade organisations or (funnily enough) agencies whose business it is to provide Performance Bonds. Bonds are normally set up for about 10% of the contract value and cost the company about 10% of the Bond value. You will normally use what are called Conditional Bonds, where stipulated conditions apply to claiming against them.

The client pays for the Bond, so it is often advisable to have the value of the Bond identified as a separate sum in the tender submission. This has some benefits:

a) A higher Bond cost will not distort an otherwise competitive bid
b) If the Bond cost is exceptionally high, the bidder (if they are the successful bidder) can be advised to seek a cheaper option
c) The Bond cost can be identified and settled, without it being buried or included as a percentage uplift amongst, for example, a mass of Schedules of Rates where it might ultimately cost the client more than the Bond's real value

A Parent Company Guarantee (PCG) can be sought where the bidder is part of a Group. It costs the bidder nothing and simply means that the parent company undertakes to stand the cost of any losses incurred by the client should the firm not complete the contract. If the parent company will not agree to provide a Guarantee, take that as a loud warning signal!

These processes and procedures relating to lowest-price tenders also apply to the financial element of MEAT evaluations, which we shall look at next.

Most Economically Advantageous Tender – MEAT

This is the most common form of tender evaluation and is most true to the ethos of achieving value for money, where 'best value' and 'cheapest' do not always mean the same thing. The only drawback with a MEAT evaluation, though, is that if you really are seeking to keep costs low, and cheap really is your priority, a MEAT evaluation will not guarantee you the lowest-priced bidder as the winner. Better quality always comes at a price.

As I have stated above, if you want a certain quality from a lowest-price bid, you will need to specify the quality – but you will still pay for it whether you evaluate it or not.

The idea of MEAT is to strike a balance between cost and quality and the relative importance of each is reflected in the declared balance of the two in arriving at the 100% bottom-line score. It is normal to have a balance of Price:Quality in a ratio of about 70%:30%, but it is not set in stone anywhere. This is procurement

– there are times when lots of rules fly around but there are even more times when there are none (and you probably wished there were). The ratio should reflect your own (or the client's) views on the relative importance of price to quality, but there are some tangible considerations that will help you arrive at a figure.

Using some exaggerated sample figures for illustrative purposes:

- If the maximum quality score were to be 5%, the quality element of the bid is unlikely to have any impact on the outcome of the evaluation as the price score will dominate unless two price bids are within a very small margin of each other. It would almost be simply a tie-breaker. The effect of this will be that price is king: tenders will be low and bidders will put little if any effort into the quality submission as they will see it has no real importance in your eyes. They may also believe the contract itself will be run along the same lines.
- If the quality rating were to be 90% instead, bidders will realise, contrary to the example above, that a high price will not necessarily lose them the contract as it only has a value of 10% of the total score. A good quality bid would probably secure them the contract at the cost to you of a premium price bid. All bidders will think in this way and you will certainly pay more for the service than you possibly need or ought.
- If the price and quality were both 50%, the quality element, again, would be quite powerful and the tendency would be for prices to creep up, albeit with a degree of caution on the part of the bidders.

Considering the logic behind the three scenarios above, you will understand that ratios in the region of 70:30 will be most effective (at least for the majority of construction-related tenders), although 60:40 is used effectively in a lot of cases. You will note that I have not said which number is quality and which is price. This is deliberate because it can be either way round: all we are trying to do is arrive at a proportion where there is an emphasis on the most important element (be it price or quality) yet the lesser element still has sufficient power to influence the outcome.

To set the above into some sort of context, tendering for a company to build a simple brick bin store may not need such a strong emphasis on quality as if you are tendering for a provider to carry out home-care services for the elderly where the quality of care has to be paramount but funds to provide the service are not limitless.

Whatever you are tendering, it is an arithmetical balancing act with no hard or fast solution. My suggestion is to create the model with the balance you think is right and try it out with other, like-minded souls. Play around by inputting figures and if you feel the balance is wrong, simply adjust the scoring to set the balance right.

Scoring Price

If your tender is a lowest-price bid process, evaluation is easy. Cheapest wins. No points, no maths: job done. Sometimes, though, you require prices for more than one element and need then to work out which is the best deal. These 'elements' *could* be set-up costs, installation costs and ongoing maintenance costs and you need to balance submissions to find the bid that presents the best overall value. The supply and installation of a heating plant might be an example of this.

How you evaluate this type of submission is a matter of choice, as these things so often are in procurement. You can:

- Ask for maintenance costs over – say – five years, add all the (three) elements together and take the cheapest overall option.
- Decide some aspects have more financial impact than others and allot each element a weighting reflecting its relative importance.

You may ask why some aspects of a financial bid would be more important than others, when it all has to be paid anyway? This is a good question, and it applies to elements of the quality aspect as well, but a myriad factors often prevail and need to be taken into account. One reason I can think of as an example of why different financial aspects may need to be weighted differently is, where the spend would be a mix of capital and revenue money, you may wish to minimise the revenue costs – perhaps there is a predicted shortage of revenue income next year, and you have capital money that has to be spent before the end of this. On this basis, you could put a higher weighting on the revenue element of the evaluation model, encouraging bidders to keep this figure low.

Let's look in more detail at weighting and assessing price bids.

Weighting Price Bid Elements

Taking our example above of revenue versus capital spend, we can work through a simple weighted evaluation exercise, assuming a tender where we have Revenue and Capital elements weighted in the ratio of 70% to 30% respectively. How you arrive at these proportions is a matter for your judgement. Later on in this chapter we will look at determining weightings for quality scores and you can apply that same logic to price bids. For now, assume that the logic has given us the 7:3 weighting we are going to play with now.

To make life easy, we will ignore any quality scores for now and allow 100% of the tender marks for price. This is therefore like a lowest-price bid, but slightly more complex. Let us imagine there are three bidders and they have submitted the values shown in the table in Figure 7.1 below. We need to drive revenue costs down, so have weighted the scoring in the ratio 70:30 marks for revenue:capital respectively.

Contractor	Revenue bid Value 70%	Capital Bid Value 30%
Smith (S)	£1,200,000	£650,342
Jones (J)	£1,350,000	£830,119
Wilson (W)	£1,096,350	£927,640

Figure 7.1 – Sample Price Bids for Financial Evaluation

There are various ways that you can carry out the evaluation exercise so we shall look at the simplest one here and consider the merits of other methods afterwards. Which method you use is up to you – albeit you may have to justify your choice – the only thing you *do* have to do is make it clear in your tender documentation what method you are going to use, even going so far as to provide a sample calculation if necessary. How you evaluate the bids may influence how the bids are submitted – that is part of the tendering 'game'.

The logic of this first method is that the lowest bid gets full marks and the other bids get a lower score, inversely proportional to the ratio of their bid value to the lowest bid: the higher their bid, the lower their score.

Looking at the Revenue bids first:

- Wilson had the lowest bid, so straight away scores all 70 marks.
- Smith's proportion of the possible 70 marks is calculated as inversely proportional to the size of their bid compared with Wilson's. The formula to calculate this proportion is therefore:

$$\text{'S' Score} = \frac{W \times 70}{S} = \frac{1{,}096{,}350 \times 70}{1{,}200{,}000} = 63.95 \text{ marks}[4]$$

Jones's proportion of the 70 available marks is calculated in the same way:

$$\text{'J' Score} = \frac{W \times 70}{J} = \frac{1{,}096{,}350 \times 70}{1{,}350{,}000} = 56.85 \text{ marks}$$

Note:
- The table I have used, usually drawn up on a spreadsheet, is known as an evaluation matrix.
- The providers each get a score out of the potential maximum of 70 – they do not share the 70 marks between them!

---

4. You may be best to run through these calculations yourself to aid your understanding of the arithmetic.

- Remember to always calculate the proportion using the same, lowest bid (in this case Wilson). It is easy to get mixed up.
- I have shown the scores to 2 decimal places. If the scores are really close, this may be necessary. In our example, they are far enough apart to use the nearest whole number, which we shall do at the last stage.
- You should set up a spreadsheet to do the calculation. It may take some time to do this but will more than pay that time back when the bids come in.
- The spreadsheet can also collate *all* the scores – quality and price, where you have both – to give you the final tender outcome. The benefits of this will become clear as this chapter progresses.
- In general, the formula to use for lowest price scores is:

$$\text{Bidder's score} = \frac{\text{(Lowest Bid) x (Maximum Points Available)}}{\text{(Bidder's Price)}}$$

The Evaluation Matrix now looks like this:

Contractor		Revenue 70%		Capital 30%	
		Bid	Score	Bid	Score
Smith (S)		£1,200,000	63.95	£650,342	
Jones (J)		£1,350,000	56.85	£830,119	
Wilson (W)		£1,096,350	70	£927,640	

Figure 7.2 - Sample Price Bids –Revenue Scores Added

We have started to include the scores. So far, Wilson is in the lead but they do *not* have the lowest Capital bid. We need to repeat the calculation for the Capital submissions using the same formula, remembering this time that the Maximum points available are 30. I suggest you attempt the calculations yourself before moving on to the next figure, where the new scores are shown.

Contractor		Revenue 70%		Capital 30%	
		Bid	Score	Bid	Score
Smith (S)		£1,200,000	63.95	£650,342	30
Jones (J)		£1,350,000	56.85	£830,119	23.50
Wilson (W)		£1,096,350	70	£927,640	21.03

Figure 7.3 - Sample Price Bids – Capital Scores Added

A bidder's final score is the sum of the two constituent elements, revenue and capital. This gives us the outcome shown in Figure 7.4:

Contractor		Revenue		Capital		Total Score
		Bid	Score	Bid	Score	
Smith (S)		£1,200,000	63.95	£650,342	30	93.95
Jones (J)		£1,350,000	56.85	£830,119	23.50	80.35
Wilson (W)		£1,096,350	70	£927,640	21.03	91.03

Figure 7.4 – Sample Price Bids – Total Scores Added

The picture has changed: Wilson was way ahead when the Revenue bids were evaluated, but Smith came through strongly with a very low Capital bid on the final straight and made it by just over 2 marks (2%). The gaps are, however, just large enough to round the scores to the nearest whole number: Smith scored 94, Wilson scored 91 and Jones scored 80.

This was a very close competition and illustrates the need to check – where you can – for any errors and to ensure you consider whether the lowest bid in any element is viable.

In this example, we gave Price 100% of the marks. When you have a MEAT evaluation, if Quality scores, say, 60% there will only be 40 marks available for the price. If there is only one price element, you score all bids just as we did either of the two elements above, except the maximum available score is 40. If there is more than one cost element, you divide or weight the 40 marks accordingly and adopt the same arithmetical approach as we did above. If there are three price elements, you do it the same way, but there is one more set of calculations to do. Easy-peasy.

There are – again – options and some people will score the price element differently. You *can*, if you wish, score price submissions (and quality) out of 100 (in the same way we did above) and then convert the score to the required weighted value. For example, if price has a weighted value of 40 marks, score the bids out of 100 and multiply the outcomes by 0.4. The final answer will be the same but some people find this method easier to understand.

Other Price Scoring Mechanisms

I repeat – there are other ways to score price submissions, and everyone has their favourite. The one I have covered in detail above is the simplest and the most common and is therefore recommended as the preferred model, to be used unless you have good reason to use another. I will, however, tell you of one other very different method that can be used, but mainly to illustrate the different ways that a price evaluation can be scored.

Variance from the Mean

The 'variance from the mean' method makes the assumption (rightly or wrongly) that the average of all the bids received will reasonably accurately reflect the true market value of the service at that time and that all the bids will be either above or below that figure. You therefore calculate the average of all the bids received and give *that* figure the full score – say 100%. You then calculate the *difference* of all of the other bids from that average price, ignoring whether it is higher or lower. You calculate, in the same way we did in the example above, the score value of this difference and subtract it from the 100%. This will give you the score for each bid. Sounds complicated? Maybe, but it's not, really: let's run through an example.

Picking some easy sample figures, the Evaluation Matrix would look like this:

Bidder	Bid Value	Difference from Average	Score rating of Difference	Score Rating of Bid
A	£300.00	£36.20	10.77	89.23
B	£326.00	£10.20	3.03	**96.97**
C	£428.00	£91.80	27.31	72.69
D	£221.00	£115.20	34.27	65.73
E	£406.00	£69.80	20.76	79.24
Average of Bids:	£336.20			

Figure 7.5 – Sample Price Evaluation Matrix – Deviation from Mean Method

Notes on the matrix in Figure 7.5, above:

- The maximum score possible is 100 marks or 100%[5] where the variance is zero.
- Only by pure coincidence will anyone score 100%. It would mean they actually submitted a bid equal to the average of all submissions. That would be spooky.
- You cannot calculate their score simply by using their bid value and the average bid because it does not accommodate the plus and minus values. However...
- The *Difference* from the Average figure has no plus or minus value – so it makes no difference to the logic of the method if a bid is higher or lower than the average.
- The Score Rating of the *Difference* (known as the variance) is

---

5. In this example, we can use percent scores because we are marking out of a maximum 100 points. In real life, it could be out of any value.

calculated using the principle used in the main example above, namely:

$$\text{Score Rating of } \textit{Difference} = \frac{\text{(Bid Difference)} \times 100 \ (= \text{max. points available})}{\text{(Average Value of all bids)}}$$

This calculates the difference as a percentage of the average bid. In our table 7.5, that means that Bidder B is 3.03% away from a perfect score. Had he (spookily) bid exactly the average, his *difference* would have scored zero – in this case a *good* score as a bidder wants their difference from the average bid to be small.

- Score Rating of the *Bid* is then found simply by deducting the Score rating of the Difference from the possible 100 maximum score.
- The logic is the higher (the score of) the difference, the more is deducted from the maximum possible score – in our example the 100%
- In this case, Bidder B scored the highest, with 96.97%
- In other words, their bid was 3.03% away from the average price.
- If this price element is weighted, convert the score we have, that is out of 100, to the weighted value as explained above.

There are both protagonists and antagonists of this method. I have used it but do not necessarily advocate it, but it does have its good points. Its prime value is in automatically eliminating rogue or maverick bids, be they high or low. It will rule out (or discourage) bids that are way above the average market rate and at the same time eliminate anyone trying to buy the contract. The methodology therefore means that no negotiations or clarifications are necessarily required to further assess apparently unviable bids, although an investigation would need to be undertaken to comply with the Regulations.

That is the benefit of this method. However, almost by definition, the winning price bid (even if not the final winner when quality is considered) will not be the lowest – it will always be the closest to the average and this may cause problems when recharging leaseholders through the Section 20 process (see Chapter 14).

Of course, an excessively high or excessively low bid will distort the average – this is unavoidable – but the method does soften this impact, and any bid offered has to reflect *a* market rate, albeit an inflated or a deflated one.

So long as the methodology is made clear in the documents, bidders must abide by the chosen method, and you can emphatically state, when seeking approval of any award, that the bid selected does truly reflect market values.

Scoring a Schedule of Rates

I have looked at scoring a lowest-price bid when the tender is submitted in the form of a lump sum. However, if you look back at Chapter 6 you will see that often bids are sought in the form of a Schedule of Rates or a Schedule of Standard Order Descriptions. Evaluating bids in these formats is slightly different.

At time of tender you have two options. You can:

i. issue bidders with a priced Schedule of Rates (see Chapter 6 on specifications) and ask them to tender overall uplift or discount figures against the rates you have laid down or

ii. you can leave all the price fields blank and let the bidder fill in the gaps – that is, they have to offer a price for each individual schedule.

Option (i) has many advantages: it is the quickest for the bidder to complete before returning it and is also the quickest to evaluate (for obvious reasons). The problem is, as the bidder has no control over the base prices he has to accept a swings and roundabouts situation where some rates will turn out to be generous and some will be poor – perhaps even unintentionally punitive.

Option (ii) above would be rather like pricing a Bill of Quantities (BOQ) and has its benefits: bidders can make sure they set a proper price for themselves for each element, so avoiding the swings and the roundabouts. However, it is a laborious process for them to price every item and more work for you to evaluate.

Whichever option you choose, it is normal to issue a schedule of the anticipated use of the SORs. Be sure not to guarantee that this will be the actual usage – it may be a model based on previous usage or simply an educated guess.[6] This 'model' can then be used to:

a. Guide bidders offering an uplift or discount if you are tendering using method (i) above or
b. Allow you to compare the bids submitted in method (ii) by inserting the tendered SORs into the model and evaluating it as if it were the price for a scheme (i.e. rate x usage per line then totalled). Cheapest outturn is the best cost (generally).

Which way you choose is up to you and if you are not sure, you can ask your QS – after all, they'll probably be the ones who have to do it.

As explained in (b) above, if you are evaluating a SOR with bidder-entered rates, you will need to feed the rates into your model and assess the outcome as though it were a price for a job (again akin to a BOQ). That is fairly straightforward and you evaluate the model's theoretical price as if it were a 'real' price bid.

If, however, you are asking for an uplift/discount figure against priced SORs and purely evaluating that tendered percentage, the evaluation logic has to be different. The biggest discount (or smallest uplift) wins the pricing element of course, but if you are evaluating against MEAT criteria with a quality element you need to allocate a points score to each bidder's submission to count along with the quality score and there lies an issue – if the rates laid down in the schedule are anywhere near the going market rates, tendered adjustments will be relatively close and so the returns may easily

6. Such figures or models are offered as "anticipated" or "estimates" usage.

include both uplifts *and* discounts! So why is that a problem and how do we deal with it?

You could, of course, ensure that the rates laid down are all at a marked variance from the market rate, but you still cannot guarantee that the different tenders will not include both uplifts and discounts, especially if the tender is further complicated because, say, the contract includes a range of response times.

The problem that a mixture of uplifts and discounts presents when you enter the submissions in the calculation is in the arithmetic: most spreadsheets, when assessing the points value of a bid, will not automatically recognise that -2.5% is a better offer than, say, +1%[7] and will tend to get into a bit of a mess. Let us look at an example, illustrated in Figure 7.6.

We are looking at four bidders named A to D in column 1. The red figures in column 2 show (good) tendered discounts against the priced rates and the (bad) tendered uplifts in black. Referring you to the evaluation and calculation methods described earlier when looking at lowest-price bids, if you take the best offer C at -0.3% you will allot that bidder 100 marks (column 3). However, if you then calculate the marks scored by the other bidders using the standard calculation used for priced bids earlier on, the scores for each become evaluation nonsense (see column three).

COLUMN	1	2	3	4	5
			WRONG		RIGHT
	BIDDER	DISCOUNT %	SCORE/100	20 LESS DISCOUNT	SCORE / 100
	A	-0.25	83.33	20.25	99.75
	B	2.5	-833.33	17.5	86.21
BEST OFFER	C	-0.3	100.00	20.3	100.00
	D	1.6	-533.33	18.4	90.64

Figure 7.6 – Assessment of tendered SOR Uplift/discount

---

7. Discounts are shown as negative values (because they reduce) and uplifts as positive values, hence the colour coding. Contrary to normal logic, red numbers are best in this context.

To counter this anomaly, take any reasonable number – I have chosen 20. From 20 *subtract* the tendered uplift/discount (Column 4) and then the negative (good) bids will *add* to the 20 and bad (positive) uplifts will be *deducted*[8] from the 20. Automatically, the good bids make the new number bigger[9] and a clear logic appears. Suddenly, you are assessing the more sensible numbers as they appear in column 4.

We now apply the same evaluation technique – bidder C gets 100 and the scores of the others are worked out as described earlier and we get the usable outcome shown in column 5.

Remember, the scores in column 5 show the percentage that each bidder gets of the maximum possible *weighted* score for that particular element – they are *not* their final score!

You then go on to calculate the bidders' final scores for this section in the same way as for any price bid, by multiplying the maximum possible score for the section by their percentage score achieved as in column 5. If the maximum achievable score for this price section was 30%, Bidder C would get all of it, i.e. 30%. Bidder D, on the other hand, would get 90.64% of 30%, or 27.19%.

These SOR and other price scores are then added to the quality score as part of the final stage of evaluation. It all sounds a lot of work, but the use of spreadsheets makes the job an inputting doddle once you have set the formulae up – which you will already have done, of course, to test the evaluation model.

Remember – *always* test your evaluation model, many times. It *has* to be right.

Scoring Quality

We shall look at how to set out quality questions later on in this chapter but for now shall persevere with the matter of scoring and weighting. In the section above on price we spoke of weighting different elements. We'll look now in detail on how to do this – price or quality – using the quality section as our example.

8. Standard laws of arithmetic
9. Discounts are minus values. Subtracting a minus results in it becoming a plus, thus *adding* to the 20, so bigger is better.

Setting weighting values

Let us consider an imaginary quality section and assume you have six questions: you allocate 10 points to each question and your evaluation model will have a maximum possible score of 60 marks – so far. Note that we are deliberately and completely ignoring what percentage ratio has been allocated to this part of the evaluation score (i.e. the price:quality ratio) at the moment: that comes later. Now, suppose some questions are more important than others? We are not talking about any killer or sudden-death questions here, but do you think 'experience', for example, is more important than 'level of resources'? If it is, you will need to make that question more important, i.e. worth more marks. If you think it is twice as important you can give it 20 marks instead of 10 or you can *weight* the score of that question to make it have more impact on the outcome, in this case by a factor of two. Simple.

It seems on face value easier to increase the available points, but this can make the initial evaluation harder. How can you mark one question out of ten, then the next out of, say, 30 and the next after that out of 15 and still be expected to score consistently? It also means you have to invent a new scoring model (i.e. how you assess the response) separately for each question and that can be hard work. Why make hard work? So I recommend you go for weighting and keep the maximum score per question always the same.

To set the weightings, we have six questions, A to F. You then need to go through a mental process something like this:

- "A is twice as important as B, so will need a weighting of 2 against a weighting of 1 for B. So far so good.
- Mind you, C is a little more important than A so will be weighted 2.5.
- Mind you, B is three times as important as D, so we find D will now be weighted at 1 and B at 3, with A and C increased accordingly".

I hope you're keeping up. If not, read it again, looking at Figure 7.7, overleaf).

We would then look at where E and F fit in to all of this. So far so good, but I see one possible problem: suppose an applicant scores 7/10 on question C? The weighted score would be 7 x 7.5 = 52.5 but we do not want fractions because they can get messy. We can avoid this by not having halves in our weightings: the easiest solution (but not the best) is to round up or down or (a better solution) you can double all the weightings.

So far, on the logic of the above train of thought, we have:

Question	Max. Score/ Question	Weighting	Weighted score
A	10	6	60
B	10	3	30
C	10	7.5	75
D	10	1	10
E	10	?	?
F	10	?	?
	Maximum total score		??

Figure 7.7 – Weightings Logic

For simplicity, I will say that questions E and F rate the same priority as question B (i.e. a factor of 3). Our table, adding the weighting of 3 each for E and F and then doubling to remove the 7.5, would then become as in Figure 7.8 below, where the weightings have been applied to the maximum score of 10/10:

Question	Max. Score/ Question	Old Weighting	New Weighting	Maximum Weighted score
A	10	6	12	120
B	10	3	6	60
C	10	7.5	15	150
D	10	1	2	20
E	10	3	6	60
F	10	3	6	60
	New Maximum total score			470

Figure 7.8 – Adjusted Weightings

So our evaluation model has a total possible score of 470.

Now what do we do with it?

If this were a tender return evaluation model, it would have a specific value or percentage of the total tender score – say 30% for a MEAT ratio of P:Q = 70:30 (a very common ratio). In this case, a score of 470 marks would equal 30%. What if they score 250? Easy maths:

% age score for 250 marks = $\dfrac{30 \times 250}{470}$

As a formula, this reads:

Percentage score = $\dfrac{\text{percent available for this section} \times \text{actual score}}{\text{total marks possible in this section}}$

This formula will work for any percentage proportion of the quality bid.

Combining Elements of MEAT Criteria

In the above paragraphs we have looked at declaring the models and how they work for price and quality. When working with *MEAT* evaluation models, you have to combine the two elements of price and quality. Let us look at how we do that, continuing to use the models we have drawn up above.

When combining the scores, you must have the final scores for each of price and quality *as they fit into the final evaluation matrix*. This depends on how you have set it up.

In our capital/revenue example above, the final price score was made up of 70% revenue and 30% capital. Looking again at the table overleaf:

Price Submission Scores						
		Revenue		Capital		Total
Contractor		Bid	Score	Bid	Score	Score %
Smith (S)		£1,200,000	63.95	£650,342	30	93.95
Jones (J)		£1,350,000	56.85	£830,119	23.50	80.35
Wilson (W)		£1,096,350	70	£927,640	21.03	91.03

Figure 7.9 Reminder – Scored Price Bids

We have arrived at price scores expressed as a percentage. Let us now imagine the tender was one evaluated using MEAT criteria, and that the tenderers scored the following weighted marks out of a possible 100 in the quality section:

Contractor		Quality Scores
Smith (S)		72.35
Jones (J)		81.64
Wilson (W)		75.40

Figure 7.10 – Sample Quality Scores

We will imagine that our MEAT criteria weighted cost:quality in the ratio 80:20. We now need to calculate the final scores by combining cost and quality in those ratios.

Summarising the scores, we get:

Contractor		Quality Scores	Price Scores
Smith (S)		72.35	93.95
Jones (J)		81.64	80.35
Wilson (W)		75.40	91.03

Figure 7.11 – Sample Price and Quality Scores

What these figures tell me is:

- Smith scored 72.35% of the maximum possible 20% for quality and
- 93.95% of the maximum possible 80% score for price.

Remember - no price score is 100% because, in this example, it is a combination of two separate elements (capital and revenue) and no-one was cheapest in both[10]. This enables me to work out Smith's final scores out of the maximum 20% for quality and 70% for price as follows:

Smith's final quality score is: $\frac{72.35 \times 20}{100}$ (That is, 72.35% of the maximum 20% available).

As a formula this is best written as: $\frac{72.35 \times 20}{100}$ = 14.47%

Price
Using the same logic, Smith's final price score is:
$\frac{93.95 \times 80}{100}$ = 75.16% (93.95% of 80)

Smith's final, total tender score is therefore the sum of these two:
= 89.63%

Using the same method for the other two bidders gives us an evaluation matrix with the following final scores:

Contractor		Quality Scores %	Price Scores %	Final Quality Score /20	Final Price Score /80	Overall Final Score %
Smith (S)		72.35	93.95	14.47	75.16	89.63
Jones (J)		81.64	80.35	16.32	64.28	80.60
Wilson (W)		75.40	91.03	15.08	72.82	87.90

Figure 7.12 – Sample – Final Price and Quality Scores

---

10. If we had only one price to evaluate, the cheapest *would* have achieved 100%

Smith won by a short neck: 1.73%, in fact. In such instances, the need to keep figures to 2 decimal places becomes obvious. It is not rare for keen bidders to achieve scores so close that a percentage point or two can make the difference between winning and not winning. Such a situation demonstrates you have bids that reflect true market value, which is good. It also means you have to be *extra* sure of your scoring – you need very little error to make the wrong decision. Make sure that if there is any error, you find it before the unsuccessful tenderer does. Double/treble check everything (not just the arithmetic) and ensure that any moderation exercise was properly executed and arrived at the correct outcome.

When outcomes are this close, some people advocate the use of a tie-break question or choose to declare that when bids are within a specified percentage of each other, the tendering authority will reserve the right to select the bidder submitting the lowest cost tender. Whilst I can see the merit of this approach, I don't subscribe to it: you have declared the criteria and formulae against which a bidder will be selected and, if your methodology is right, the outcome will be right – why hedge your bets? The risk of challenge is not reduced – bidders can still challenge the scores on the basis of their closeness so you are no better off, really. As I say – you pays your money and you takes your choice.

The above calculations work using 100 in the bottom line because we have used percentages in our first round of scoring (i.e. scoring the price and quality elements). However, you might not use percentages. You may have used 50 or even 270 (you can - see earlier chapters).

In general, therefore, the formula to use for calculating final scores is:

$$\frac{\text{Element Score x Overall Percentage}}{\text{Maximum Possible Score for the Element}}$$

Where:
Element means price or quality

Element Score means the score achieved by that element in the first calculation (e.g. 27/50)

Overall percentage means the percentage allotted in the price:quality ratio (e.g. 70:30)

Maximum Possible Score for the Element means the number of marks overall *for that element* in the first calculation (e.g. originally scored out of 50 marks overall). This need not be the same number for price and quality.

The final score for the bidder's complete tender is then the sum of the two final element scores – price and quality. This is best expressed as a percentage, which is what all the formulae above will give you.

For your information, there is more on tender evaluation techniques in Chapter 9 on Partnering that you may find of interest as it expands the possibilities (and so the mind) a little.

Quality – Setting the Questions and Assessing the Answers

That, in a nutshell, is the scoring process. Let us now look at setting up the quality evaluation section itself and allotting the criteria against which it is scored.

There are various legal cases that deal with issues around evaluation criteria but a landmark trial[11] between a company called Letting International and the London Borough of Newham served to lay down some key legal considerations regarding the detail of evaluation criteria, their use and the scoring of them.

These considerations cover the evaluation of questionnaires *and* the evaluation of tenders so the following four key points are very important. They are:

1. You *cannot* ask any question in a tender that you have already asked in the Qualification Questionnaire.

---

11. Much of EU tendering procedure has been set in place by court decisions. Chapter 14 discusses the impact of precedents.

2. You must always make clear the criteria you are going to use to score the submissions.
3. You must show the scoring method, and how points are awarded against each element in each question.
4. You must score in the manner you have declared.

Looking at each of these in turn:

1. **You *cannot* ask any question in a tender that you have already asked in the Qualification questionnaire**

   This means that sometimes you have to decide whether the information you are asking for should be in the Questionnaire as a fact about the business and its operation or in the tender as a matter relating to the proposed contract that you would like to score competitively. The level of a firm's resources might be an example – are you interested in what their level of resources is as a fact (i.e. qualifies them to bid), or are you keen to give a company a higher tender score if they have or will commit to putting more staff on the road to meet your contract's demands?

   That is not to say you cannot ask questions on a similar *topic* in the questionnaire and in the tender and this is sometimes a way round the dilemma – you simply have to be careful to make sure that you are definitely looking at different aspects of the issue.

   In the example of resources, you could ask at questionnaire stage how they recruit, what qualifications the operatives have as a minimum, staff structure and so on. At tender stage, you could ask what specific resources are going to be allocated to your contract, their qualifications and experience and how they will be managed to ensure quality of service. This would be quite OK.

   A key aspect of such questions – whether in the Questionnaire or the tender – is that what they declare becomes contractual upon award. They submit responses as a matter of fact, sign them off as such, and so this becomes a commitment. Often, questions are asked at Questionnaire or tender stage primarily to achieve a contractual undertaking from the bidder. The tender or Pre-Qualification documentation should make this point regarding contractual commitment quite clear.

2. **You must always make clear the criteria against which you are going to score the submissions**
This is probably the area at highest risk of challenge, and great care needs to be taken when proceeding through this potential minefield. In Chapter 15 I talk of legal precedents, and these have created some very interesting discussions around the subject of quality evaluation criteria. As you know, I have tended to avoid too much 'hard' legal stuff in this book, but I will give you a short legal rundown on two cases that are relevant here.

In one legal case (Letting International v. LB Newham, 2008) the courts agreed that the plaintiff (Letting) had not been fully (and so properly) informed of the quality criteria and sub-criteria against which their bid would be assessed. Letting claimed that, had they known, they would have structured their submission differently, scored a higher mark and so may well have won the tender. The judge agreed, found in favour of Letting, and consequently sent shivers down the spines of local authority procurement specialists everywhere: from then on, they would have to tell bidders absolutely everything, in the finest detail, which they were using to score the bids. (As I refer to later, it suddenly felt like they were virtually going to have to tell the bidders what to write).

Despite the reticence, this cautious, heart-on-sleeve approach became the procurement profession's norm, but a re-think may now be necessary. In June 2010, as a result of another, similar challenge (Varney v. Hertfordshire CC), the judge found in favour of the defendant (Herts. CC), explaining that such a level of disclosure (of criteria) that the plaintiff claimed should have been given would have placed an undue burden of disclosure on the contracting authority (my words)[12].

12. Details of both these cases – and others – can be found through most good internet search engines.

So there you have it – two cases and two precedents, both on the same topic and both quite different. You pays your money and you takes your choice and, when the next challenge on criteria goes to court, the lawyers concerned will be avidly reading the small print of each judgement in an effort to find the line of distinction between the two cases and so determine on which side their respective clients' positions lie. As the saying goes – it's probably not over yet.

On this basis, I urge you to do two things: (1) keep abreast of court cases and (2) err on the side of caution until this seeming discrepancy in the law's position is resolved (if it ever is). In the meantime, I shall press on without apology in a way that minimises, for you, the risk of challenge.

Consider the scenario: you are providing feedback to an unsuccessful bidder for a £5million contract. The bidder is disappointed and annoyed that they have not won. You explain to them that they scored badly in Question 9 because they did not explain that they would ensure that all operatives wore company uniform and the bidders turn round to you and say "but you did not ask about that." You have a problem. There must be no doubt about what you are asking a firm to tell you, and what you are looking for in terms of standards in quality or quantity.

The point is fair – if you cannot say what you are looking for, how can they be expected to know? The industry is becoming much more challenge-orientated and the risks of a challenge are much higher; since the Remedies Directive came in, the potential impact of a lost challenge has become disastrous (see Chapter 11).

Because of this, when you set a question, you need to be quite specific and list *all* the points against which you are going to assess the response and how you are going to score them.. To some, the question always arises 'are we not just telling them the answer?' and in a way, yes, but that need not be a problem. We shall look carefully at this conundrum as we go along.

To begin with, you need to consider the format in which you want your responses submitted, and this can vary even within one tender evaluation model to suit the question. You can provide a tick-box set-up for answers that are numeric (i.e. "tick box 1-5 or box 6-10 or box 11-15", etc.) or for a simple yes/no. You can provide boxes or spaces for one-word or numerical responses (e.g. "Please give your Accreditation Registration number"). Where the response requires script (e.g. a method statements) you can provide adequate space or ask for them to be clearly titled and enclosed as a separate document. Make sure you specify a maximum word count and state if you allow the inclusion of photographs, diagrams, tables, etc. Be sure to allow enough scope for the information you require, but at the same time remember that you have to read *all* of the responses so enough should be the maximum.

To look at how specific your explanation of assessment criteria needs to be, let us consider a tender for a contract to carry out extensive internal refurbishment work to an occupied block of flats where access will be needed for internal repairs to each of all the properties. Your quality element of the tender will include an assessment of a bidder's process for gaining entry to occupied premises, which you obviously want to be in line with your customer care philosophy.

Your question might be: "Describe your procedures to gain prolonged access to occupied premises in order to carry out internal repair works".

In describing your assessment or evaluation criteria, you might say: "Bidders must describe their methods for gaining access to premises".

Marks for you out of 10 for this? Probably zero. They may be able to describe a procedure, but is it any good? Will it deal with the demographic mix of your block? If you think it is not good enough because it does not deal with the deaf couple in number 67, bad luck on you because you have not said it has to. All they have to do to get full marks is describe their methods: *that* is all you have asked for.

In describing your assessment criteria, you could say:
"The method statement must clearly show how the bidder will manage access issues relating to residents who are:

- at work all day
- disabled
- unable to communicate in English as a first language
- unresponsive to requests for access"

(There may be more criteria, but this is an example).

Marks for you out of 10 for this one? Creeping up – a good 6 or 7, I'd say. You have now covered the key problems you know exist within the demographic mix of your residents. However, you have not said how you will judge whether the proposed method for dealing with each issue is good or bad. Let's try and get it still better:

> "The method statement must clearly show how the bidder will manage access issues relating to residents who are:
> - at work all day
> - disabled
> - unable to communicate in English as a first language
> - unresponsive to requests for access"
>
> The method statement must demonstrate a timely, sympathetic and co-operative approach to the various needs of different residents when seeking access to their homes and show how access will be assured for the extent of time required to complete the necessary works. The procedure should make use of a Tenant Liaison Officer or other suitably placed person and include the advance distribution of printed information for residents, designed to allay any concerns residents might have about giving contractors access to their home."

Marks for you out of 10 now? Certainly a good 9 – I cannot believe a perfect 10 is attainable.

If you were writing it, you might well phrase it differently or pick up on different key issues that you would like to see covered, but note how the level of detail means you have asked for what you want to see in the procedure. You have also made clear the standard of service you expect within the procedure and yet have still not given them the actual answer or told them what to write.

You could also make use of phrases such as:

- "The method statement must demonstrate ..."
- "Explain how the contractor will deal with..."
- "Consider issues relating to..."
- "... give evidence of..."
- "... show experience of..."
- And so on.

Subjectivity and taste

These last bullets are also useful when you encounter the need for an evaluation model that depends on taste rather than hard delivery facts – such as in a design competition. This can be tricky because, in design, beauty is in the eye of the beholder so how can you say 'A' is definitely better than 'B' and justify the outcome to those who lost?

To deal with this, you need to first of all lay down hard criteria where you can: for example, the building's function – how many desks it must accommodate, ground floor public access, open space to the rear, disabled parking, etc. – and these can be hard-scored. The *design* (i.e. what it *looks* like), however, is different, and you will have to explain that the scoring of the less tangible elements of the proposal will, for example, be based on "the subjective views of the evaluation panel". You can then revert to phrases like those bulleted above to give an idea of what (mainly in principle) the evaluation panel will be looking for.

This may be interesting, because you will find it likely that the panel members may all be looking for something different. They will probably all have different opinions and the final decision will be reached at a Moderation Meeting (see later) where you will identify the winner and define why, so that feedback to the unsuccessful entrants is comprehensive. Try and be positive – remember no design (what it looks like) can be called emphatically wrong because it all hangs on taste. Not meeting planning requirements may be a hard negative fact, but any such requirements should have been made clear at the outset. The best way to feed back, then, is to emphasise why the successful submission won rather than saying what you did not like about the others. This will avoid counter-arguments and maybe give you an easier time on a difficult day.

Setting these criteria out takes time and thought but, as I have said, it is well worth it to [a] get the right provider on board and [b] keep you out of jail[13].

3. **You must show the scoring method, and how points are awarded against each element in each question**

I earlier explained in detail the overarching method for scoring questions and the use of weightings. Here I shall look at the issue of sub-criteria – another key area of risk of challenge. You may have allocated 10 points to a question, but regardless of how many points you allot, you have to say how that total is made up by any constituent parts.

Consider the question:

"Describe in no more than 200 words your approach to working in occupied premises, including:
a. making access arrangements with tenants
b. managing works with tenants in occupation and
c. maintaining essential services during the course of each day
d. dealing with complaints regarding work carried out in a person's home."

Stating that this question has a value of 10 marks and a weighting of, say, three it is assumed that the marks are scored evenly across all four aspects of the question – in other words 2½ marks each. If this is not the case and one aspect is worth more marks than another, *you have to make this clear*.

---

13. Not literally – but maybe out of court

Two points on this scenario will help to illustrate some points about setting up questions.

i.   2½ marks per answer to me sounds ludicrous. In practice I would not allow this situation to prevail (as I have said – avoid half marks). To overcome it I would engineer a fifth part (e), split one of the other four into two or divide the four into two separate questions, giving five marks each. There is always a way.

ii. The sample questions I have given require a written response so my immediate thought, on re-reading the question, is that [a] we need to state a maximum word count for each (remember that?) and [b] 2½ marks for, say, a 200 word response is also a nonsense both in terms of reward for the applicant's effort and the assessor's ability to score.

To develop this train of thought, let us look at the 9/10 example I used in (2) above. It said:

"Bidders must clearly show how they will manage access issues relating to residents who are:
- at work all day
- disabled
- unable to communicate in English as a first language
- unresponsive to requests for access

The method statement must demonstrate a timely, sympathetic and cooperative approach to the various needs of different residents when seeking access to their homes and show how access will be assured for the extent of time required to complete the necessary works. The procedure should make use of a Tenant Liaison Officer or other suitably placed person and include the distribution of advance printed information for residents, designed to allay any concerns residents might have about giving contractors access to their home."

Imagine you are going to score this question out of 20 marks: you can do this in one of two ways. The first is to state that the question as a whole is worth 20 marks and you will then make an overall assessment of the response and rate it out of 20. This is OK up to a point but there are some issues.

The question will be scored by more than one person. Remember – always have an evaluation *panel* so that scores can be compared – you do not want the view of one person: you want a considered score that accurately reflects the quality of the response. If two people disagree on the score a discussion will ensue to determine the right outcome. If an assessor says "I knocked off three points because they did not mention the Tenant Liaison Officer", who said the TLO is worth 3 points? This point gets even stickier when an unsuccessful bidder asks the same question. In this particular instance you will literally have no answer that would stand up to a challenge.

An even worse scenario would be where you had told bidders that "This question is worth 20 marks" and then, when asked by the scorers you advise them to knock off 3 points if they fail to mention a TLO. All of a sudden you have *undeclared sub criteria* and that is not legal. It is *the* cardinal sin of tender evaluation (at least according to the Letting International v LB Newham judgement).

To avoid challenge, always show how the marks for a question will be made up. There may be instances where a written response has a points score full stop, but that will be rare. In most cases the question will require coverage of more than one point and in such a case, each point needs its share of the score identified.

The example question above is not set out in a way best suited to showing how the total score is made up. You could separately explain this by bulleting the criteria underneath the question and allocating points but this makes the whole thing a bit bulky. A better way is to re-format the advice section on criteria in such a way that point values can be included within it. One such format might be:

Criteria	Score value
Bidders' responses must clearly demonstrate that their access procedure:	
1. Provides for residents who are:	
• at work all day	1
• disabled	1
• unable to communicate in English as a first language	1
• unresponsive to requests for access	1
2. Has a timely, sympathetic and cooperative approach to the various needs of different residents when seeking access to their homes	5
3. Assures access for the extent of time required to complete the necessary works.	3
4. Makes use of a Tenant Liaison Officer or other suitably placed person and	5
5. Includes the distribution in advance of printed information for residents, designed to allay any concerns tat residents might have about allowing contractors access to their home."	3
Total marks for this question :	20

Figure 7.13 – Example of Scoring Sub-Criteria

Notes on Figure 7.13:

- This is a sample format – you can make it what you want.
- You could break any of the criteria down further and make the scoring model more detailed to suit. The criterion about the TLO is a key one that could be expanded to include processes that they would carry out, qualifications, experience, etc. Or…
- The TLO may be featured in more detail in a separate question,

if the role is felt to be that important which, on a project of this nature, it probably would be.

- Note that not all criteria have the same score value – I have made some issues more important than others.
- 20 marks for this question is regardless of any additional weighting applied as part of the overall evaluation methodology.
- My question is worth 20 points because that is what I have made it worth. I could have allocated points to the criteria and then seen what the total came out to – it may have been 35. This would not matter but it is easier for people to score if all questions have the same value (albeit with weightings applied later on) and if the score is a natural number. 10, 20 and 50 are good scores to use, if you can, because they suit peoples' natural feelings about marking models. 100 is also comfortable for people, but it can get a bit clumsy.

By way of reinforcement and to broaden your understanding, I will look at another example.

Imagine you are going to tender a term contract for the day-to-day maintenance and repair of boilers in 450 homes on a 24/7 basis, you know that certain levels of resources will be necessary, with transport suitable for carrying tools and spare parts. Already, a question is taking shape – you know that a firm comprising one man and two labourers cannot cope. Big firms could cope; really big firms could cope easily. All of a sudden we see a scoring pattern developing:

Using the above example, let us consider some fictitious scoring.

Scoring out of 10:

<10 qualified heating engineers on staff	- 0 points
10 – 20 qualified heating engineers on staff	- 4 points
21 – 40 qualified heating engineers on staff	- 7 points
41 – 60+ qualified heating engineers on staff	- 10 points

Not a perfect question, not a perfect scoring framework, but it is a start – and that is how it works. The process is not instant, so don't believe it is. Each question has to be carefully considered and, when you have the question, you need to think of the answer (or level of capability) you are looking for and the scores for a range of responses dependent upon (a) your needs and (b) the service's tolerance levels. In the above example, if you do not see any benefit in a firm having over 40 staff, then do not give applicants more marks for having a resource of over 40. You will need to think this through, with the assistance of people in your organisation who will know these things – your 'experts'.

Go back a paragraph or two – a one-man-business could not cope: Looking at the model above, we seem to have settled on a minimum of 10 operatives to cope with the service[14], albeit only just (and so the lowest score). But we have here an opportunity to include a sudden-death question:

Please state how many fully-qualified FTE[15] heating engineers you employ.

(Please note: Any firm with less than 10 full time engineers will not proceed any further in the tender process).

As we are dealing with gas boiler servicing and repair, another could be:

Please enter your *GAS SAFE* Registration number

(Please note: Any firm unable to provide this information will not proceed any further in the tender process).

Because without this legally-required registration, they could not legally provide the service.

---

14. Made up number – do not take this as fact!
15. Fte = Full Time Equivalent – standard staff-size measurement.

Such sudden-death questions are a very useful way of quickly weeding out no-hoper submissions and, hopefully, of advising firms who do not stand a chance of winning not to apply in the first place. You can then, at time of evaluation, go to these questions first in order to quickly reduce the number of questionnaires (both Corporate and Technical) that have to be fully assessed. If you have more than one such question, put them all together in a special section at the beginning, to make the weeding-out quicker and easier.

Evaluating Resources

The new Regulations now allow you to evaluate the CVs of the staff who would be allocated to your contract, although there is no guidance on how this should be done. This can be tricky: how do you rate or score a person who is recently well-qualified against a person with lower qualifications but 20 years' experience? Bearing in mind you may have to justify your marking, I offer the following points for guidance on what you may wish to consider. You can:

1. Specify minimum qualifications for the roles you are looking at
2. Request a minimum period of practical experience
3. Explain that you will balance qualifications against experience and hopefully...
4. Explain *how* you will balance these two criteria.
5. Request details of (a set number) of previous similar contracts they have worked on
6. Make it clear that CVs will be scored based on how closely, *in the opinion of the evaluation panel,* the person meet the organisation's requirements.

Point 6 is akin to the issue of subjectivity in the evaluation of design proposals (see Design Competitions), where it is best to declare that an assessment may be, in part or in whole, subjective. Challenges cannot be raised on the basis of an evaluator's *opinion* – only for them not sticking to the rules.

Here is an example of how you might do this in practice:

> "Please provide details of three recent projects similar in size, scope and complexity to the one being tendered and explain this person's rôle in that project (no more than 250 words per project). You must include the title of the project and client contact details in order that the information supplied may be verified if necessary."

Firstly – note the word limit. If you do not set a limit, you will end up with War and Peace to read for each submission Also, do not specify "no more than one side of A4", otherwise you will still get 800 words, but in Font 6. You can also ask for pictures, location maps, diagrams, etc., if you think they will help.

Where were we? Oh yes – scoring. The question above is one where the scorer needs to make a value judgement on the match or comparative qualities of the example submitted to the tender or project in hand. If we stick to a 10 points maximum we could guide the scorers (and the bidders) thus:

No matching qualities	0 points
Similar in type but smaller and less complex	1 to 3 points
Similar type and size, less complex	4 to 7 points
Similar type and size and matching or exceeding complexity	8 to 10 points

In this case, the scorer has a range of options in each grade to exercise their judgement *and* the applicant knows how the scoring is being done. This is easiest for both parties. You can add that a score of zero will warrant the submission ineligible for further consideration on the grounds of non-compliance and remove them from any further assessment.

You could elaborate the question by going on to ask for details of successful outcomes and so on but you will need to be careful of the context: in a tender you can only consider the CV in relation to your project (you cannot look back – remember?). Therefore, the way you word the question is very important or *you* could be the one who is non-compliant!

In general, always make sure there is a clear explanation (i.e. a warning) about any pass/fail scoring mechanism you may have in your evaluation model.

Points

Although I have tended towards a maximum score of 10 in the examples above, mainly for ease of illustration, scoring each question on a basis of 1 to 5 is favoured in many institutions, with the scores allocated on a basis along these lines:

SCORE	RESPONSE
0	No response submitted or an inadequate response meeting none of the criteria.
1	Poor answer meeting some of the criteria but weak in some areas.
2	Reasonable response, meeting most of the criteria.
3	Good response meeting all of the criteria.
4	Very good response meeting all the criteria and exceeding some of them.
5	Excellent response exceeding all the criteria.

Bidders should note that a score of zero on any question will eliminate them from proceeding any further in the evaluation process.

The actual definitions may vary, but you will almost certainly encounter this evaluation methodology. In principle, this model is fine but it gives the scorer little leeway as you cannot give half marks[16]. Another problem with 1 to 5 is that between each score (1 and 2, for example) lies a margin of 20%, and, to me, 20% intervals in a scoring model is a big gap: "it is better than a 2 but not a 3, but I cannot give ½ marks - I *have* to be 10% wrong whichever way I go." Not my favoured method, as I say, but it is used in a lot of places so you need to know about it.

We have looked at simple scoring, above, but it can get complicated so I will try and make it easy for you. Firstly, do not get hung up on 'easy' numbers. People like bottom-line scores to total 100 or 1,000; well – good for them, but to me this is a waste of time (and brain) – the score will total what it totals, and that's it.

You need to make each *question* easy to score, so I suggest you allocate them 10 points each. People find it easy and natural to score out of 10.

If we go to another extreme and give, say, a maximum of 50 marks per question, this will give *lots* of leeway; it is up to you, but don't get carried away. Also, it is easier if you can give the same number of points to each question as this makes it easier for the scorers or evaluators to work with.

4. **You must score in the manner you have declared**

This is the simplest one of the four key rules to explain: having told the market how you are going to score their submissions, do *not* change it. Simple as that. If you do need to change it (say you have discovered an error), you will have to go out to all bidders, advise them of the changes and allow them time to submit amended bids to suit the new model. Believe me, it is easier to make sure you get it right in the first place and stick to it – design the model and test it until it breaks so you *know* it is fit for purpose.

---

16. Half marks – you can give half marks, but why would you? How is 2½ out of 5 easier than 5/10?

Moderation

Lowest-price tender evaluation is an arithmetical exercise and the only cross-checking required is to make sure the mathematics is correct. It is always wise to have the sums checked – errors in arithmetic do occur and if spreadsheets are being used, the sheet will do the sums it is instructed to do but sometimes the instructions are wrong. One minor error in a formula can lead to a major mistake in the outcome.

With quality, there is much more leeway – even with clear criteria there is always room for subjectivity, interpretation and opinion on how well a requirement is addressed. In addition, you will find one person will consistently mark high and another low. That is not important so long as the trend is commonly applied across all submissions. What is important is that all the views are brought together at the end to arrive at a common view on the outcome.

When you have got all evaluators' scores in, collate them onto one sheet and hold a Moderation Meeting where the final scores for each question are agreed. *Do not* average the scores: a score may be low because the evaluator missed a point when reading the response and the average will be unfairly skewed.

At the Moderation Meeting, the Chair (preferably and by rights someone who was not involved in the evaluation) goes around the room and seeks agreement on each question's score. Discussions will take place as different views on the answers are aired and a common, agreed view will be settled on. Normally, this is an easy process and the persistently high or persistently low markers will recognise this in themselves and agree on a method of moderation that makes it possible to score all bids fairly across the board.

If it is not possible to agree a score, it may be necessary to re-visit the submission and literally assess each evaluator's views on the response. This can take a little while and so sufficient time should be allowed for this when setting up the meeting.

The scores are not final until everyone is happy with the final scores, agreeing that they properly reflect the outcome of the tender process.

Post Tender Interviews

Post-Tender Interviews (PTIs) are a very useful implement in your evaluation toolbox. Whilst they have little (if any) value in a standard, run-of-the-mill project or lowest-price tender, they are particularly valuable when tendering larger or more complex contracts (a Partnering or a long measured term contract for example).

PTIs have two primary purposes:
1. Seek clarification of a tenderer's bid
2. Posing additional questions for evaluation

Looking at these in turn:

1. Seeking clarification of a tenderer's bid is exactly what it says – an opportunity for you to confirm that the tendered sum is correct and to clear up any uncertainties or anomalies that have arisen during the process of the tender evaluation. These meetings allow members of the evaluation panel to clear up any ambiguities or vagaries in a bidder's submission and firm up their score. Such clarifications can always be raised with bidders in writing, of course, but in some instances, particularly for quality questions, interviews are considered preferable.

2. Posing additional questions for evaluation means just that: bidders are asked to present and respond to questions on subjects or aspects not covered by the written submission part of the process. These presentations will, of course, be scored separately and additionally to the written submissions.

It is not unusual for the award of a contract to depend entirely on the outcome of the post-tender interviews and for this reason the whole process must be taken very seriously, be thoroughly prepared and conducted in a proper and above-board manner. We shall look at this manner in some detail.

When seeking clarifications (of a bid), you must be careful that this process does nothing more than just that: you need to ensure that a bidder's score is not enhanced overall simply because, upon meeting them, they have become a preferred bidder.

The panel must also be careful that a bidder does not win at interview stage purely on the basis of their honed presentation skills, rather than because of the *substance* of what they have presented. The fact that you need to advise bidders in advance what the questions are going to be makes this more difficult. Firms can get very slick and, once invited for interview, can gear themselves up by rehearsing for the day using selected staff, all with the intent of securing themselves everlasting fame and an Oscar for their interview performance. This is not what you want and needs to be avoided.

Firstly, to avoid being set upon by a cohort of directors and sales reps, make sure when you invite firms to interview that you specify that those attending must be the people who will be operating the contract if the bid is successful. You will know who these people are because you will have asked for their details as part of their written submissions.

Secondly, if the presentation is to be on a more mundane and factual (i.e. boring) topic such as the system to be used for contract monitoring or audit, added gloss is hard to apply. If, however, you are asking for details of the bidders' approach to something more esoteric and inspiring, the opportunity is ripe for fluff and shine. To avoid this, you can advise bidders that the interview questions will be selected from a given list of several. You can even avoid advising bidders exactly *what* you will be asking on each topic if your wording is clever enough. You can choose the questions you wish to ask from the list at any time: often this can be done just before the interviews, when the written submissions may have alerted you to an area you would like to explore. The main thing to remember, though, is that you must ask all presenting bidders the same questions, score them against the same criteria, and give them the same amount of time each to present: the even playing field again.

With the ITT you will need to provide either the scores for each individual potential question or explain a more general scoring mechanism that you will apply across the board, to all questions. This method is akin to the subjective scoring methodology I have discussed earlier. For example, you can award scores on the basis of answers that "demonstrate a thorough understanding of..." or "most closely match the needs of..." and so on. It comes with practice.

The questions do not all have to have the same score value: they can each be scored on the basis of their individual importance to the contract. All you have to do to compensate for this is ask questions whose scores add up to the total value allocated to the interview process. The panel can score every question out of 10 or 50 (say) if you like to make it easier, and weightings can be applied, just as with written submissions, to each question and to the overall interview score.

Post Tender Interviews can be very valuable but are at risk of challenge and so must be conducted properly. There are key points that need to be observed to minimise this risk and that I shall cover them here.

- Firstly, bidders must be advised in the Instructions to Tender that Post Tender Interviews will be held as part of the tender process.
- You need to advise them what format the interviews will take, albeit an agenda is not required at this stage
- You should also advise bidders, as accurately as you can, when the interviews are scheduled to take place so that they can leave space in their diaries.
- You can advise them what questions will be asked at interview stage (if they get that far) or you can advise them that bidders will be advised of topics when invited for interview. This does not prevent you advising on the overall score for the interview stage, although you will not be able to advise on any detail: this will have to come with the advice on the questions – see later.
- The score-value of any part of the PTI must be included in the tender evaluation criteria guidance and, of course, made an integral part of your evaluation model and matrix.

- When you actually invite firms to interview you need to give them plenty of notice, be clear on date, time and venue, provide an agenda and ask if any special provisions need to be made (for example, disabled facilities). Explain what IT provision will be made available.
- Do not make the interviews too long – it is not a war of attrition. An hour and a half is normally about right. This is not a rule, it is my view.
- An hour and a half will allow you – if you wish – to fit three interviews into a day but do not rush the process in order to achieve this – it is not a target. Always allow for interviews to over-run.
- Avoid bidders crossing paths on the day – this can be awkward and not in keeping with the principle of bidder confidentiality.
- Do not spread the interview stage over too many days. Evaluation fatigue will set in and the evaluations will not be consistent.
- Ask that a copy of any presentations be made available for the panel
- Make sure the same interview questions are put to all bidders, although supplementary questions can be bidder-specific.

The interview panel and the interviews

You need to establish a panel for the interviews, made up of relevant persons. You do not need to include the whole evaluation panel but you may do so if you wish and if it is not too many people. However, the same people must be present at every interview.

If a presentation on, say, design is part of the requirements, Member representatives might wish to be there – or there might even be a special post-tender presentation session for Members. You may also want resident representatives to be involved. Remember, though, that because of their position, Council Members may not score – they can only have opinions.

Book the panel members' diaries well in advance to avoid them being on leave or otherwise engaged when you need them. It will be a lot of people to secure in one room at specific times.

Book the meeting room well in advance as well. There seems to be a growing worldwide shortage of meeting rooms and if this comes to bear when you want to hold your interviews you will be in trouble. Ensure the room is adequate in size and that the layout can facilitate a presentation to the panel. Provide water, other refreshments and IT facilities as appropriate and make sure that the reception desk where the bidders are to arrive know they are coming and what to do with them.

Always have a pre-meet with the panel. Explain the agenda, the protocol of the meeting, what questions will be asked and who will ask them. In particular, make sure everyone understands the scoring model and their rôle in the interview process. You do not want to appear unprofessional or have any glitches caused by the panel being disorganised.

Conduct the meeting to the agenda and to time. Set clear time limits on any presentations and use a guillotine system – do not allow over-runs because this is unfair to other bidders and will wreck your schedule.

Once an interview is over and the bidder has left, you should, as a panel, discuss the outcome and agree or moderate your scores for that session. As each post-interview discussion takes place, it is reasonable to re-visit earlier interview scores, just as you would when assessing written submissions, in an effort to ensure the scores across all interviews are consistent and fair and reflect the relative merits of each. The mop-up after the last interview is particularly salient in this regard. You can look at the overall outcome and this will save a further moderation meeting just to combine the interview scores with the written tender scores and come to a conclusion.

Make sure you allow for these sessions in your meetings schedule.

Which companies?

Deciding which companies to invite needs to be done carefully and of course in accordance with the process you have described in the documentation. It is not normal to ask every bidder to interview -

that does not make logistical sense. You must, however, be careful that the outcome of the interview cannot impact on the tender scores in such a way that a firm *not* invited to interview could have won if their interview had scored well.

To do this, and disregarding any adjustments that post-tender clarifications might produce (although you cannot disregard this really, of course), if the post tender interviews are worth 15 points, you need to invite any bidder where 15 points added to their current score could cause them to win the bid.

If, as I said above, you are asking for clarifications as well (which you almost certainly will be), you may need to increase this theoretical 15-mark margin to allow for the scores of bidders' written submissions to be increased on any questions where you have a query.

In real life, if you do not invite a firm who *could* win, they will feel disadvantaged and may challenge. If you invite a firm who could not possibly win – they will not be happy either when they find out (at feedback) and, if really miffed, may seek costs for a wasted afternoon. Put simply, do not invite anyone who cannot possibly win the tender.

Lastly, whilst you should not ask to interview any more firms than you need, do not invite less than three. In similar vein to aiming for no less than three bids to evaluate, this makes the logistics as easy to manage as possible whilst giving you benchmark interviews to assist with your assessments.

Remember – to really get to the nub, face-to-face questions are really effective: interviews enable you to do this.

The Evaluation process

It is worth noting that the methods explained here can be equally applied to a tender or the contract-specific elements of a questionnaire, with the exception that questionnaires have to look back at capability and capacity, and not offer proposals for your specific contract, going forwards.

Letting International were able to successfully demonstrate that an evaluation was not in line with the methodology described in the documents, and so they won. So, when setting questions, establishing model answers, defining evaluation criteria and setting a scoring structure, just bear in mind you *may* just have to go to court, stand in the dock and justify the outcome of your process. This can be a sobering thought.

A sound quality evaluation process depends upon having a good evaluation panel. You need to have panel members who are knowledgeable about the questions they are scoring and you should have at least three score each question. Their scores may differ – that's fine: moderation will take care of that.

The scoring must be based purely on the submission, not on any personal views or prior knowledge and quality scorers should not know the price submissions in case it influences their marks for the quality element.

If using an e-tendering portal, scoring can be done directly onto the screen, either at the desk or even at home. If it is hard-copy, it is best to have all the submissions in a room and have the panel spend the day on it (or longer if necessary), working on prepared scoring sheets you will have provided. The panel members must all make notes explaining the rationale behind their scores, and these notes should be in some detail: they will form the basis of your Standstill (Alcatel) letters, they may have to support a debriefing session and they may even have to be submitted as evidence in court to defend against a challenge (!).

At a moderation meeting (which may be on the same day as the scoring) the panel will discuss why individual scores are different and arrive at a consensus score for each question – do *not* average – and in this way end up with a final score that is the considered decision of the whole panel.

The tender manager (the person running the tender process – it may be the 'project leader' and will probably be you) will ensure that all submissions are available for the evaluation panel.

Whilst confidentiality with regards to Questionnaires is not an important issue, tenders *must* be secured, distributed and maintained in a blatantly confidential manner. The higher the profile, the more elaborate this must be seen to be. Tenders should ideally be maintained in a locked room (if not secure within an e-tendering system) where the evaluation takes place, and all evaluation notes kept with them, secure from the outside world.

For lower-value tenders, submissions may be copied and distributed or e-mailed, but always the issue of confidentiality must be stressed. It is not uncommon to ask members of an evaluation panel to sign a declaration stating that they have no conflict of interest issues and will maintain strictest confidence. How far you go with this is a matter of your judgement and the normal regime of your organisation.

It is better to condense the evaluation process into as short a period as possible. This not only keeps the programme under control but, more importantly, helps ensure a degree of consistency in the evaluations. All evaluators should be aware of the deadline for completion of the process and some people like to make the evaluation a day's event (or more than a day if it is a big tender).

Except in the most extreme of circumstances, the panel must be kept the same throughout any tender process in order to maintain scoring consistency. To achieve this, you will need to check panel members' availability as part of the timetabling process.

The tender manager must ensure marking sheets, suitable for the purpose, are available in a format that will facilitate the collation of the individual scores at the end. You may have one sheet per bidder or a set of sheets to cover all bids in one 'book'. This sheet should be clearly marked with the evaluator's name and have room for a scorer's notes that briefly explain how the score was arrived at. This is not to be an essay but must be legible and comprehensible in case the suitability of a score or a view is called into question. These notes are invaluable at the Moderation meetings. This sheet should be clearly marked with the evaluator's name and the bid/s it relates to.

Once the evaluation process is complete, *all* the paperwork is kept for the records. The winning tenderer's paperwork is kept as part of the contract documents and the unsuccessful bids are stored – with the evaluators' notes – in archive for a suitable period of time.

What is 'a suitable period of time'? This depends on the views of your organisation, but essentially they need to be kept until they are no longer needed, i.e. until such time as a challenge on the tender process is no longer likely or viable. This is probably around a year, but there are schools of thought that feel such documents ought to be kept for a period in line with the Statute of Limitations – six years. I think this is excessive, and suggest you seek the view of the organisation's Legal Section.

The winning contractor's papers need to be kept, along with all paperwork relating to the actual contract's operation, in line with the Statute of Limitations, i.e. six years, counting from the last action of that contract, which would normally be the final settlement. With construction projects, this last date could be the date the final account is settled upon expiry of the defects liability period.

If the contract is sealed as a Deed by your legal section, the Statute of Limitations period is extended to twelve years and this is done where there is a need for an extended period of contractual liability. This extended period of contractual liability is normally required when a building is constructed or some other lasting, tangible legacy comes out of the contract. Sometimes, illogically, contracts are sealed or not according to their monetary value.

Summary

- Tenders can be evaluated using either lowest price or MEAT evaluation criteria
- Use weightings to reflect the relative importance of different aspects of the evaluation model
- Schedule of Rates and Standard Order Descriptions tenders require slightly different evaluation techniques
- MEAT means **M**ost **E**conomically **A**dvantageous **T**ender
- With a MEAT evaluation model, the price:quality split must be declared

- All evaluation criteria must be made clear within the tender documentation
- The marking or scoring model must also be made clear
- Once declared, these cannot be changed
- Post tender interviews can be a very valuable tool in a tender evaluation process
- An evaluation panel's scores must be moderated, not averaged, to achieve the final score
- Ensure evaluators keep notes explaining the rationale behind their scores
- Remember: the details of your evaluation process can be asked for if you are challenged so care needs to be taken; be sure to check all scores and outcomes and keep all paperwork, including individuals' notes.

8 // Frameworks and Approved Lists

What are they?

A Framework may at first seem to be just the same as an Approved List, and sometimes it can be very similar. The fundamental difference between a Framework and an Approved List is that the former is a proper contract (or Agreement) and the second one isn't – it is simply a list.

Because a Framework is set up as a contract, if it has been tendered in the appropriate manner (see later) it can also be used as a mechanism for letting contracts that are above the EU threshold without actually having to go through a full EU process. This is perhaps the most useful and distinguishing aspect of a Framework and an Approved List cannot trump this.

Frameworks are the current procurement fashion item and are becoming more and more prevalent, so beware: the phrase 'tender a Framework' rolls very easily off peoples' tongues (some see it as the panacea of all procurement ills) but you need to be sure that a Framework really is the best tool to achieve what you are looking for and this debate must be part of your procurement strategy process. Having said all that about Frameworks, I shall nevertheless take a look at Approved Lists first and then move on to consider Frameworks in more detail.

Approved Lists

An Approved List is a list of providers, all of whom are deemed eligible to work for an organisation as a result of some form of vetting procedure. Approved Lists can be set up for specific areas of work (e.g. construction) whilst some organisations insist they have Lists covering all service requirements.

An Approved List will have categories that further refine the areas of supply for which a company is considered eligible to work and companies can apply to be in more than one category. It is not unusual for categories to then be further divided into contract value ranges so that companies [a] will only be invited to bid for contracts that suit their financial profile and [b] will only be in competition with others of a similar size.

The vetting process normally assesses a firm's eligibility against corporate criteria; in other words, such lists merely tell you which firms are *suitable* to work for your organisation. For obvious reasons, they cannot offer any guidance on competitiveness or quality regarding your specific contract, and exactly what corporate criteria are assessed as part of the vetting or approval process depends on the organisation.

It is a usual requirement, although this depends on an organisation's Contract Standing Orders, for any competitive process below the EU threshold to be conducted through an Approved List where there is a suitable category. This is often the undertaking an organisation will give in return for firms suffering the application process. Contracts that approach their EU threshold *have* to be publicly tendered using an EU process: firms on an Approved List may take part in the tender, of course, but their position on the list gives them no material advantage, and no shortcut through the selection process.

To get onto a list and become approved, a firm has to go through an application process. It is sensible (but not always the case) if the Application Form is a mirror image of the Corporate PQQ, in other words, same document – different cover. After all, why should eligibility and suitability to work for an organisation be different just because you are on an Approved List?

Approved Lists must have rules that make it clear how they operate: how the application process works, how firms are selected to bid, what will get a firm suspended and how they can appeal against such a decision. The normal process for selection of firms for a tender or quotation is by rotation, based on when they were last asked to bid; an organisation can have variations on a theme but this is the usual process.

The operation of a list must also have processes to ensure that approved firms remain eligible: they should always have up to date insurances, accounts and other documentation necessary to meet the organisation's approval criteria – the assumption is that any firm on a list is always known to be eligible to work.

It is important that performance records are kept of listed firms; scores against performance indicators should be entered during and at the end of each contract. It can then be assured that – so far as is possible – firms who are on the list have a proven track record which can be referred-to should any performance issues arise.

Approved Lists may be open or closed. Open lists allow companies to apply at any time and closed lists are only open for applications at certain, specified times. Regardless, all lists must be periodically reviewed – every five years is common – and at this stage all firms on the List have to re-apply and new firms get the chance to seek approval. Such a complete re-let is a major exercise but it can be eased by opening the List one or two categories at a time.

For all these reasons, Approved Lists require a resource to operate them and, ideally, a suitable IT system to maintain the records and even perhaps perform the rotation process.

There are bespoke IT systems that can be used for running approved lists (often part of an e-tendering software package) or your organisation may have a bespoke system of their own; there are also companies that operate Approved Lists on behalf of client organisations, and these earn fees by charging firms who apply.

The detail of Approved List set-up and operation depends entirely on the client organisation and its approach to procurement so I will not go any deeper into the subject here. I have provided information such that you will understand any list that your organisation may operate, but be aware that if you are ever asked to set up an Approved List from scratch, you will certainly need to do some more research.

Frameworks

Frameworks differ from Approved Lists in several respects:

- They are contracts or 'Framework Agreements'.
- They must comprise either one or three or more member firms (i.e. they cannot have two).
- They are populated through a competitive process (i.e. a full-blown tender) not by application.
- Once a Framework has been set up, no more companies can be added.
- Frameworks can last no more than four years.
- You do not *have* to use a Framework once it is set up and you guarantee member firms no level of work at all.
- You cannot rotate firms on a Framework but...
- You can award contracts through a mini-tender, cascade or call-off process (all described later).
- Frameworks can be used for the procurement of above-threshold contracts without going through a full EU tendering process (see later).
- If tendered appropriately, other organisations can use your Framework and you can use theirs (see later).

This is a lot to take on board, so I will go through each bullet point in turn.

Frameworks are contracts or 'Framework Agreements'
Approved Lists, whilst they imply mutual commitments comprising eligibility to work and opportunities to tender, there is no tendering process to secure a place on a list and no actual, signed contract in place. Lists merely identify firms who have applied and are known to be suitable in certain respects to work for the client, and any contractual arrangement can only be struck after a further, full procurement process.

Whilst you also require a process (of some sort) to commission a service from a company on a Framework, firms have to go through a competitive process to be awarded a place on the Framework in the first place and there is still an Agreement in place that lays down strict terms and conditions relating to the Framework, its operation and the company's position on it.

One point – some people refer to a 'Framework Agreement' and some to a 'Framework Contract'. So long as they are both referring to the overarching legal agreement relating to a firm's place on a Framework, they are both talking about the same thing.

Frameworks can only comprise one or more than two member firms

This is an EU ruling that is designed to prevent collusion between two providers 'in competition' in a Framework. There is a logic, albeit somewhat limited as my personal view is that, if the intent is there, three can collude just as well as two, but there you are.

There is no maximum number of firms you can appoint, but you will need to balance the number of providers against potential demand. Six to eight firms is common.

Why only appoint one? The logic of this lies in the other bullet points – once appointed you do not have to use the firm but on the other hand other (named) client bodies can – so the Framework gives you much greater flexibility than other contract set-ups can.

Frameworks are populated through a competitive process

Frameworks are populated through a tendering process. Because of the nature of Frameworks, this is usually a full EU procurement and will normally use MEAT evaluation criteria. As firms on a Framework are likely to be used for a variety of contracts, when tendering a Construction Framework (i.e. a Framework to be used for tendering construction projects) it is most common to tender either against a Schedule of Rates or against a model (i.e. pretend) project. Schedules of Rates are most useful when this is to be the method for selecting providers (see later) whilst the 'model' method is best if selection will be via mini-tender. For a Consultancy Framework, you could tender hourly rates or fee percentages, (or both!).

Once a Framework has been set up, no more companies can be added

Unlike an Approved List, once the award decision has been made and the Framework established, no more companies can be added – you cannot have an *open* Framework. Firms on a Framework can drop out or be suspended, however – although in reality they simply cease to be used. Only when a Framework is re-tendered can other companies try to get on it.

Frameworks can last no more than four years

Frameworks can be established for any period of time up to four years, but they cannot exceed that limit. Personally, I see little point in setting them up for less than four years because, if you do not want it after, say, three years, you can simply stop using it.

The Regulations do allow for a Framework to exceed four years provided there are extenuating circumstances that will convince the EU that there is really no option. In my view – good luck with that!

Remember, though – any contract set up *through* a Framework can run its full course – it does not have to suddenly come to an end when the Framework itself runs out.

You do not have to use a Framework once it is set up

As I have said above, you have no legal obligation to use a Framework and can always go out to public tender, even for contracts that the Framework could serve. It is part of the deal that, even though part of a Framework, a company has no guarantee of any work from it whatsoever.

You cannot rotate firms on a Framework

There are various ways you can procure firms off a Framework, but rotation is not one of them. Rotation is reserved for other set-ups, such as Approved Lists. There are three key ways in which you can select a company from a Framework:

You can award contracts through a mini-tender, cascade or call-off process.

Mini-tender: properly called a mini-competition, it is just what it says: you invite firms to bid in a tender process, and the 'only' purpose served by the Framework is the initial, pre-selection stage. It is called a mini-competition, but some Frameworks are such that the contracts let through mini-competitions might be worth many (and I mean many) millions of pounds.

Cascade: Framework member firms are listed in order (usually of tender score – highest score top) and you use the top-ranked firm first to provide the service until they reach capacity, when you 'cascade' the opportunity down to the next one on the list. If the top firm re-gains some capacity, the cascade process starts back with them again. The order of firms in a cascade set-up will be decided by their points score (or price) at Framework tender stage. What I have done in the past is to call this points score their 'Supplier Rating' and have this Rating adjusted through a performance management regime, increasing or decreasing it in direct proportion to their performance against set KPIs[1].An optimum target of, say, 85% is laid down whereby 85% will have no impact on a firm's rating. Any score above or below this optimum will impact on their rating – up or down – proportionately.

This potential change of rating values – and hence the pecking order in the Framework Cascade process – will sharpen firms' attention to performance and allow better performers the chance to rise up the ladder; those performing less well will tend to fall. This also helps deal with a company scoring highest at time of tender remaining at the top of the list despite declining performance.

1. KPIs – Key Performance Indicators – see Chapter 17.

Call-off process: This is when firms have tendered against a Schedule of Rates (or it could be item prices in a catalogue), and these rates stand throughout the life of the Framework. The client then calls off their requirements from the firm that is offering the best prices for their specific needs. You can often negotiate the published Framework Rates down, but avoid playing one provider off against another as that constitutes a mini-competition and a mini-competition is not a call-off.

Once the provider is selected, a contract is struck in line with the stipulations of the Framework Agreement and Protocol. We shall look how you set up Frameworks in such a way that they can meet the demands of these commissioning processes a little later on.

Frameworks can be used for the procurement of above-threshold contracts without going through a full EU tendering process.
This is considered to be the biggest benefit of Frameworks. If a Framework is tendered appropriately, using a full EU process, then you can subsequently tender contracts that are above the EU threshold, through the Framework, without going through the whole EU procedure. This obviates the need for OJEU Notices and any pre-selection process other than one you might wish to include or that the Framework Protocol lays down.

This brings time and cost benefits to any procurement *but* the Framework *must* have been procured in a way that will allow this to happen namely, the commissioned subsequent contract must fall within the category or categories specified in the Framework's OJEU (remember how I emphasised issues around the scope of a contract as specified within an OJEU Notice?).

Other organisations can use your Framework and you can use theirs.
This is another key benefit of a properly-tendered Framework. I say properly in the context that other organisations can use your Framework as if it were their own provided they have been specifically named in the OJEU as a potential user. It used to be

that you could specify, for example, "all London Boroughs" and that would suffice but there have since been some court cases (there always are) which have established a precedent whereby each potential client must be named individually (yes – that would mean all 32 London Boroughs, by their full name). This anomaly can be overcome by quoting and providing the link to a website that legitimately lists all the organisations making up the required list (e.g. the London Councils website for all London Local Authorities.

Conversely, your organisation can use another client's Framework provided it has been set up in such a way that you can, i.e. 'you' have been named in the OJEU. There are several Frameworks set up specifically with a view to allowing a variety of clients to make use of them. The Crown Commercial Service's main *raison d'être* is setting up such Frameworks (but they are mainly for supplies and services). There are, however, several construction Frameworks that have been set up by Regional Improvement and Efficiency Centres and other organisations such as Central Purchasing Bodies[2] that Public Bodies can make use of[3].

When considering using one of these Frameworks, it is essential that you:

i) First determine that you can legally use it. You should check the OJEU as this is key. Search through SIMAP or your legal team will help you, if necessary. You can ask the Framework 'owner' themselves, of course, but due diligence would recommend you check eligibility for yourself – the 2015 Regulations put this onus on you.

ii) Secondly, you need to check the terms and conditions the Framework uses to see whether or not they correspond adequately with those of your own organisation and whether they can be varied if they don't. If these checks give a green light, the next step is to

iii) Find out how the Framework operates (i.e. how firms from

---

2. These bodies are set up specifically for this purpose: they must not be 'commercial' – they must only charge fees that cover costs.

3. I cannot name available Frameworks here as this could be perceived either as a recommendation of those I include or as discriminatory against those I do not. Your own research will be necessary. Sorry!

it are selected) to confirm that the process is suitable for you. An example query might be, "Does it allow quality assessment as part of the mini-competition?".

When using another organisation's Framework, costs ensue. It is normal for the authority that tendered the Framework to engineer an income from its use to cover the costs of tendering and maintaining the Framework and sometimes even to make a profit. This income is generated either through charging the user directly or through levying a percentage of the tender sum on any company securing a commission through it.

Frameworks have 'constitutions' and 'protocols', set out at time of tender, that lay down all the details of how the Framework will run: in other words, its rules. These cannot be varied so, if the rules do not suit, you will have to go elsewhere.

Tendering Frameworks

When tendering a Framework, all of the issues discussed above will need to be considered within your strategy: you need to decide:
- what it will cover (scope)?, therefore...
- will it have multiple panels (categories)?
- who (if anyone) will you invite to use it – and at what cost to them?
- how will it operate? Therefore...
- what will your evaluation model comprise? and ...
- how many firms will you aim to have on it?
- and so on.

This is quite a lot to think about. A Framework can start off simple and get complicated. Avoid this. If you are asked to tender a Framework to provide a particular service, stick to that and beware of 'advisers' bolting on other things to take advantage of the tender process. That having been said, it is not uncommon for a Framework to be tendered covering a range of categories, known in Framework language as Panels. A good example is construction consultancy services where a Framework might have separate Panels for Architecture, Quantity Surveying, Building Surveying, Planning Consultancy and so on.

Each of these Panels would operate in the same manner, and client bodies would tender within or select practices from the Panel that is most suited to their needs. You cannot tender one contract across more than one Panel. You can tender different contracts for one project across different Panels.

You can invite or allow any client body you wish to use your Framework, but they must be specifically named in the Contract or OJEU Notice. It is best to ask them in advance if they wish to be included in the OJEU. The benefits from having other clients listed is that competition will be keener amongst providers if the potential return is greater, and the organising body (in our case, you) can earn money from this usage, as explained above.

All information regarding the set-up and operation of the Framework will be in the documentation referred to as the Framework's Constitution or Protocol. It will state the number of Panels, the intended number of operators (providers) and so on. It will also state, if using mini-competitions, how many providers must be invited to bid – often it will say all of them. Of course, this will not apply to Cascade or Call-off.

For the benefit of user clients, the evaluation model used for the framework needs to be made known within the Framework's literature. This element is key to how the Framework will be used:

- If using mini-competitions to commission contracts then often a tender against a model project may be best, and quality may or may not be a notable percentage of the score.
- If using call-off, then you will need to tender using a schedule of rates and this makes it more difficult to include a quality element. If quality is hard to include, you can specify the quality standards you require. For some types of contract this will be fine, for example a straightforward supply Framework.
- If using a Cascade commissioning method, you could use

price only but the model does lend itself beautifully to MEAT evaluation criteria to arrive at your 'Supplier Rating'.

You need to decide on a suitable number of firms to meet the potential demands of the contracts arising from the Framework. On occasions you may decide on just one firm, and we looked at the benefits of such a Framework earlier, but the circumstances where this would be the most suitable setup are relatively few and far between.

You will therefore nearly always be looking at awarding a place on the Framework to three or more companies: most Frameworks end up with about six or eight providers but there are no rules on how to make this decision – it is simply a judgement call. Points to consider might include:

- What might your level of demand be?
- How many organisations might (actively) use the Framework?
- Simple call-offs for items – say office machinery – may require fewer providers as a client's needs are quickly met.
- Construction Frameworks will involve longer-term commitments of greater value so more providers will be required.
- If cascading, will firms near the top of the ladder reach capacity quickly? This may happen if you successfully encourage smaller firms, with limited capacity, to bid for the Framework.
- If one Panel is covering a range of requirements (rather than a Framework with several, distinct Panels) you may require more firms to cover the range of services the Framework makes available.

It should now be clear to you that a Framework Contract or Agreement is primarily a *structure* with a range of legally-enabled flexibilities: it can be a versatile tool if used (and set up) correctly but must not be seen as the way all procurement should go. Always consider other options, which we have discussed elsewhere and which include term contracts and the use of Lots and Packages.

Summary

- Frameworks are a contractual structure that gives the client flexibility in the procurement of services.
- A Framework can have more than one Panel, to meet the needs of different disciplines or categories.
- If the Framework is appropriately tendered, it can be used for above-EU threshold tenders without going through the whole EU process.
- A Framework can only last for a maximum of four years but contracts awarded *through* a Framework can continue until their proper conclusion.
- Contracts can be awarded off a framework through call-off, cascade or mini-competition.
- You can use another organisation's Framework provided it has been set up appropriately and the terms and conditions are suitable.

9 // Partnering

What is a Partnering Contract?

Arising from the recommendations of the Latham[1] and Egan[2] Reports, an approach towards the management of primarily construction contracts was devised that advocated a co-operative approach rather than the adversarial approach hitherto accepted as normal practice.

The Government encourages the use of a collaborative approach to contracts and – whilst the concept has its friends and its enemies – it is incumbent upon Local Authorities to be seen to be driving this initiative forwards.

The approach is called Partnering – which is not the same as a partnership. Legal, commercial Partnerships can be formed, but that is not the context in which this guidance is set. Whilst the relationship is based on a contract, the Partnering principle lays down that client and contractor – and the supply chain as well if it is done properly – work together as partners to deliver a common aim to the mutual benefit of all parties.

Do not get this close relationship between client and provider confused with PPP or PFI contracts.

PPP and PFI Contracts

PPP stands for Public Private Partnership and PFI stands for Private Finance Initiative. Both are, in essence, the same rose by a different name in that they both use private (or commercial) investment to fund the provision of a public service.

The key principle is that the private sector provides the up-front capital investment for perhaps an airport terminal, a new bridge, an academy or a hospital and then operates the associated service. It is usual (though not essential) for the provider to construct

1. Sir James Latham: "Building the Team", 1994
2. Sir John Egan "Rethinking Construction" 1998

the asset (e.g. the building) and the value of the contract is often therefore correspondingly high – hundreds of millions of pounds. The private partner recoups costs and profits over the life of the contract, which is usually 20 or 30 years or more.

PPP and PFI contracts have not been an unbounded success, either in operational or financial terms, and are treated with more scepticism now than when the idea was first mooted many years ago as the solution to restricted public sector investment.

The procurement of a PPP or PFI Contract is a gargantuan task and it is unlikely that you will be asked to tender one, at least in the near future, but you should know what they are.

Why Partnering?

The British construction industry was blighted with time and cost overruns, variable quality and ensuing litigation. The Latham and Egan Reports independently concluded that a more collaborative approach was required between client and providers on the basis that a successful project should be a mutual goal.

It was considered that a contract operated on the basis of shared profit alongside shared risk would achieve this joint buy-in to making a contract succeed.

Partnering depends primarily on trust between all parties and such trust cannot be contracted into. Partnering will not work at once but has to grow and develop over the life of the contract. For this reason, although many projects are partnered, most Partnering contracts are for longer terms to allow this development to happen and the benefits to be reaped.

One key aspect of partnering is the opportunity for added value. These are particularly feasible in longer-term relationships, and in this context Partnering can see the generation of community benefits, local labour training schemes and so on. Proposals for such benefits can form a part of the tender evaluation process but beware of ending up evaluating what amounts to 'added value'.

How do Partnering Contracts operate?

Forms of Contract and Agreements

The contract form used as the basis for Partnering is not critical – although some might say it is. Many people choose to use a Standard Form (GC Works, JCT) with a Partnering Agreement overlay. Some people advocate the use of one of the forms of contract specifically designed for a Partnering arrangement and there are now a few of these. The PPC (Project Partnering Contract) was designed by solicitors Trowers and Hamlins to fill this gap in the market, followed quickly by their TPC Term Partnering Contract. There is also a good Partnering option within the NEC Suites[3] and JCT have a 'collaborative' option, but it is beyond the scope of this Guidance to discuss the relative merits of each.

If you are not using a Partnering contract form, a Partnering Agreement needs to be drawn up by all parties to overlay the contract. A Partnering Agreement lays down in clear and non-legal terms how the 'Partnership' (note the inverted commas) works and operates on a daily basis. It covers all key issues such as payments, audit, low-level dispute resolution, conduct of meetings, profit share and so on.

It is inherent in Partnering that all decisions are mutual. For this reason, most of the vital elements of the relationship are agreed after award at a series of meetings held specifically to set up the Partnering Agreement. The more that can be determined at this stage on the way the contract will work, the more the principles of Partnering are being applied.

As well as confirming all the 'official' details (names of partners, roles, etc) this is the working tool of the contract and it can even lay down methods and processes that are different to the original contract – providing all parties agree. If the Agreement fails, remember – it is not a contract variation and so the founding contract is king.

---

3. This is not an advert for either – these are cited as examples. See Chapters 4 and 9 for more information on Forms of Contract

The Partnering Agreement may well be changed as the relationship develops and this is normal as trust builds up. Again, so long as everyone is in agreement, this is fine.

Some Partnering arrangements have an Agreement that simply lays down the higher-level principles of the relationship. In these cases, there is an additional document – the *Operation or Operating Manual* – that sets out all the day-to-day operational 'stuff'. This method works perfectly well and has its advocates, as it separates aims and aspirations from the hands-on elements of the contract. On the other hand, it is yet another document that will require updating each time there is any change in structure or process.

So long as the Partnering works, the Agreement will serve to run the contract through to its conclusion. Normally, if you need to revert to the base contract itself to sort something out, the Partnering element has died.

If a 'proper' partnering Form of Contract is used, a Partnering Agreement should not be necessary but if you feel it can add value you may wish to draw something up over and above the basic contract; this would ostensibly create a set of 'unofficial' amendments (i.e. changes to practice).

It was also common practice for all of a Partnership's key stakeholders to sign up to a 'Partnering Charter'. This Charter comprised a statement of (good) intent, where all stakeholders, at an inaugural workshop, agreed the Partnership's key aims and objectives and composed a mission statement, etc and had a ceremonial sign-up by all concerned to a large certificate or document fit for public display. This ceremonial approach is less common nowadays as partnership contracts attract less fanfare, but such an event can have a positive effect and give the necessary profile where the project involves residents.

The Tendering Process is often more complex for Partnering than for normal contracts as the bidders have to demonstrate their attitude towards the principle of partnering and submit their bid in a way that matches the approach. Additionally, there are more variables that need to be assessed.

Sharing the pain

A founding principle of partnering is the establishment – so far as is possible – of the client and the provider as parties on an equal footing, with a common goal. On this basis, both parties sign up to a contract and / or agreement that states clearly that risks will be shared and so will any gains or losses (losses are the pain bit).

It needs to be remembered that in a successful partnering arrangement, any and every problem or issue is *everyone's* problem or issue – not "the contractor's" or "the client's". Solutions are worked out as a team and it is on this principle that any gains or losses are shared.

How these gains and losses are apportioned needs to be decided as part of the procurement strategy and, as the client, you will be deciding a lot of this before the contract goes out to tender. Above all, though, you need to strike a balance that will serve as an incentive to the contractor and yet be fair to all, ensuring that you can still demonstrate value for money at the end of the day.

Agreed Maximum Price

Cash gains and losses are shared between the provider and the client in proportions laid down in the contract but this begs the question of how do you determine what these gains or losses are? Financial gains or losses are measured against a cost agreed at the start. If the contract is a project, this will be the tendered sum and is called the 'Agreed Maximum Price' (AMP)[4]. Once the project is finished, the final account sum is compared with the AMP and the difference constitutes the gain or loss.

If the contract is not a project but a term contract for, say, responsive repairs, the AMP is arrived at by pricing the work using the Schedules of Rates costs agreed at time of tender. Gains or

---

4. Agreed Maximum Price: sometimes referred to as Target Cost or Guaranteed Maximum Price.

losses against the *actual* cost of the work can then be calculated but it is normal – for obvious reasons – to do this periodically (say every month or every three months) when the contractor's invoice for that period is compared with the costs as ordered on the system.

Of course, a combination of these two processes can be used and any type of contract using a schedule of rates can be evaluated periodically, at the end of a phase or at the conclusion of the project. As always – so long as the methodology is known, and agreed to and stuck with.

Open book accounting

The next question of course following on from this is that of financial integrity, trust and openness. Even if the levels of trust and honesty are beyond any shadow of doubt, the fact is you are spending public money, there is accountability and the financial probity of the contract and the use of this money has to be shown. To achieve this, it is normal for Partnering Contracts to operate on an open-book system where the provider's accounts are open to view by the client so that their honesty and integrity is proven. There are cynics who say contactors have two sets of accounts – the 'open' set and the real set that you, the client, will not see. I have (a little) more faith and maintain that, whilst this may, on occasion, be the case it is generally *not* the case. In reality, if you feel this, there is probably something wrong in the partnership; if you find out this is the case, then there is *definitely* something wrong. Partnering is built on trust and integrity and generally, if this is what you give out, it is what you will get back.

Overhead and Profit

To ensure probity, it is common for the accounts to be regularly audited by the client and for the figures to be checked. You can include in this audit a 'partnering' audit whereby the client can check not necessarily the actual figures (as a QS or auditor would) but the way the contract has been operated by the provider to affirm that the principles of partnering have been maintained. An example of where this may particularly apply is when, for example, a task or a supply item is not covered by the original specification or the SORs and needs sourcing from scratch. Under these circumstances the contractor would normally seek competitive quotes for the requirement, select the best option (probably using MEAT criteria) and supply it to you with an added sum on top for what is called Overheads and Profit. This is normal, and the Overheads and Profit figure (known as OHP) is normally tendered as part of their bid.

That brings us on to tendering.

Tendering Partnering Contracts

As you can tell from all the above, what I said earlier is true – tendering partnering contracts is more complex than most other types because there are many more angles to consider if you are to select the right partnering contractor. I shall therefore take time to look at the key aspects of tendering a partnering contract.

Form of contract

Deciding on the Form of Contract is no different here than for any other contract so I refer you to Chapter 4. If you are not sure, seek views from those who have made these decisions – be they right or wrong – in the past. Nowadays, an online search may bring some light to the matter but primarily I would speak with your QS, who may well have had experience of this before and who could advise. Whatever your choice – have a reason.

Tendering to make it work

If you get a good partner on board, provided you do it right, the partnering element will work. For this reason, the quality element of your tender needs to weigh heavily on the bidders' attitude towards and experience of partnering.

Make sure you seek references from clients with whom they have partnered in the past and ask pertinent questions regarding the bidders' attitude, honesty and approach; how easily were differences of opinion settled? And so on. Of course, you will want to know that they achieved more gains than pains, but essentially a partnering contract will work if the participants' *attitude* is right. If it is, cash gains will (normally) follow.

It is a good idea to have post-tender interviews (PTIs) for partnering contracts. At these interviews, you can ask for a presentation on the bidders' experience of partnering elsewhere (to support their bid) and their approach towards partnering with you, and then ask various probing questions. All the while you will be doing 'normal' human appraisal things like observing body language and – as I call it – 'seeing the whites of their eyes'. You should get a sense of how well you could work with them as a team and use the scoring mechanism of the PTI to help ensure you get the right provider on board. Remember, with PTIs, you need to advise bidders in the tender documents of the way they will be conducted, what will be required and, if the interviews are to be scored, how.

Sharing pain and gain

I spoke earlier of sharing gains and losses. At time of tender you have two choices, really:
i) Stipulate how gains and losses will be shared between the contractor and the client, or
ii) Ask bidders to propose splits as part of their tender.

This latter is by far the more complicated, and the normal way is to specify the split as in option (i). Your only task then is to make a decision on what the split is to be. Remember – it does not have to be the same percentage whatever the figure. You can vary the proportions to maximise the stick and carrot effect.

For example, if the split is 50:50 for the first, say, 20% of savings and for the next 15% the contractor will reap 10% against your 90%, the contractor may feel that an extra 10% of the additional 15% of savings is not worth the effort. He will probably be right – making savings takes time and effort and time and effort all cost money. The same principle applies to the proposal of shared losses. There are no rules on this: do what I always say, put some figures together and play with them – see what works.

The second option – having bidders tender gain and loss share – is more complex but it means that the better proposal can figure in your selection process. If you do choose this route, what ranges and weightings you apportion in this section of the evaluation is still up to you, just as it is with all the other aspects.

How can you tender this pain/gain share aspect? You need to have a model that will take into account gains and losses, and each of these over a range of values. Here is a sample evaluation matrix. To save space, I am looking only at shared gains to begin with; shared losses require a mirror-image of this as I shall explain afterwards.

Remember – I am not advocating this model: I am simply showing you how I arrived at it so that you can see the logic and, if you like it, you can consider it (modified as much as you wish) for yourself another day.

Row ID	TENDERED SPLIT OF GAINS - CLIENT SHARE						
A	1	2	3	3	2	1	
B	0-2%	>2% to 3%	>3% to 4.5%	>4.5% to 6%	>6% to 7.5%	>7.5%	**Score**
C	85	80	40	40	60	60	110.83
D	85	80	75	75	40	40	135.83
E	95	85	80	80	75	70	160.83
F	85	75	60	60	55	50	125.83

Figure 9.1 – Sample Gain Share Evaluation Matrix

Looking at each row and column in turn:

Row A	These are the weightings (i.e. 1 to 3) for the score in each anticipated range of gains. Assuming a standard bell-curve distribution, most gains would be in the region between 3% and 6%.
Row B	These are the ranges I have selected for the scale of the gains: 0-2% saving, over 2% to 3%, etc. These are *purely arbitrary figures* that I have chosen.
Rows C to F	These are the bidders' submissions for the *client's* share of the gain in that range. To illustrate, The bidder in row E is offering to pay the client 85% of any savings of between >2% to 3%. Generous, I know, but that is what he is like. The bidder in row C is offering 80% to the client. (Not probable figures, but used to test the model). These figures will constitute the score for each bidder, and will be weighted according to the figure in row A.
Column 'Score'	In my example, the score is achieved by multiplying each tendered share by the weighting in that range, adding them together then dividing by 6 (the number of ranges) to give the average score per range. This achieves: a) the best overall share for the client and b) Tendered shares which will become contractual upon appointment.
All	The outcome. Remember – what we are looking for here is the largest number, because they are client gains.

Remember – this is for the gains element. You will then have another matrix just the same for the losses share, when you will be looking for the lowest figure.

When you have these figures, you carry out the last stage of the evaluation as follows:

<u>For the gains</u>: Award the highest score 100%. In our example above, that would be Bidder Row E. So 160.83 = 100% and you need to assess each other bid proportionally as explained in Chapter 7. Bidder Row C has a score of 110.83 so their percentage score would be:

$$\frac{110.83 \times 100}{160.83} = 68.21\%$$

This means that Bidder Row C has scored 68.21% *of the weighting for that element.* If the weighting for that element is 10% of the whole price element of the tender, then Bidder Row C has scored 68.21% of that 10%. Final score: 6.82, to be added to their other elements' scores.

<u>For the losses</u>: You use exactly the same process, but this time the *lowest* score has 100%, because the lowest share of the losses is in the client's favour.

Using another similar example:

Row ID	TENDERED SPLIT OF LOSSES - CLIENT SHARE						
A	1	2	3	3	2	1	
B	0-2%	>2% to 3%	>3% to 4.5%	>4.5% to 6%	>6% to 7.5%	>7.5%	**Score**
C	30	30	25	30	10	5	46.67
D	50	50	45	45	40	35	89.17
E	50	50	50	50	50	50	100.00
F	35	35	40	40	40	35	76.67

Figure 9.2 – Sample Loss Share Evaluation Matrix

The functions of each row are the same as for the gains, and the Bidders C to F are the same people that submitted the gains bids above.

The *only* difference is, this time, they are tendering the client's share of any *losses*. This in turn means that, this time, we are seeking the *lowest* bid, using the same methodology we used to calculate the score for the gains share.

From the table above, the lowest bid is from Bidder C who scores just 46.67 marks, which we will allocate 100%. We then need to assess each other bid, so that the higher their score, the smaller their score as a percentage.

Using bidder D as our example, Bidder D's score percent is:

$$\frac{46.67 \times 100}{89.17} = 52.34\%$$

Again – remember – this is 52.34% of the maximum score for that section. If we allocated 10% for that section, whilst C got 100% of 10%, i.e. 10%, D achieved 52.34% of 10%, which is 5.23%.

The arithmetical detail here is not new – it was covered in chapter 7 for lowest-price bids. Although this is not price, the same principle applies – simply comparing numerical bids with the lowest (or highest) offer.

This evaluation process sounds complex, and in a way it is. But as I said – if you want to ask the bidders to propose these pain/gain shares as part of their tender it will involve some complexity.

Overhead and profit

OHP is much easier and is evaluated like any other numerical bid. The lowest figure is in the client's best interests so award the lowest OHP figure 100% of the weighting and calculate each of the others' scores accordingly, just as I have done above.

Main price bid

If the tender is for a project, then the financial evaluation is just as I have explained in Chapter 7 and the winning bidder's tender sum becomes the AMP.

When tendering a Schedule of Rates, it can get more complicated (as usual, I hear you say) and you have two options: bidders can price the schedules or you can price the schedules and bidders tender an uplift or discount percentage as they see fit. Both of these methods are covered in detail in Chapter 7 and I refer you there.

The only additional comment I would make is that you may also wish to have a set of Standard Order Descriptions (SODs – see Chapter 6). These, instead of describing individual tasks, describe complete small works. My example in Chapter 6 is kitchens: a SOR may describe removing a wall cupboard, assembling the new unit, hanging the new wall cupboard, fitting the door, etc all as individual tasks with a set price for labour and material. A SOD on the other hand will describe the whole process as one, complete job. Order the job under a SOD reference and the operative will turn up and remove the old unit and fit a new one all against one item. If you have SODs as well as SORs, the evaluation will be the same for both, but each will have its own share of the evaluation marks to be allotted.

Final evaluation

The Quality element of a partnering contract tender will attract a high proportion of the marks – remember, the success of the contract will depend on the right partner being appointed. The price element will comprise the total of all the sub-section scores bidders have achieved in each of the different cost aspects you adopt. In my example above, that will be scores for main contract price, pain/gain shares and OHP. How you weight each of these aspects is up to you and again this has been covered in Chapter 7.

Partnering after award

Regardless of trust, it has to be (openly) demonstrated that the Partnership is operating with probity and integrity. For this reason it is incumbent on all parties to ensure that rigid controls and audit procedures are in place and that there are regular reports to the

Partnering Board (those responsible for running it on a daily basis) which are made available for all other interested parties – such as residents and Members. Similarly, payments and profit-sharing exercises must be processed in a proper and formal manner, as they would be with any conventional contract, so as to demonstrate compliance with all due financial processes.

Such minimum requirements will be laid down by the client at time of tender as they have a duty of care of the public purse. The Partnership may afterwards feel it useful to increase the range of auditing and reporting in the pursuit of greater efficiency but the established level of financial control must never be compromised.

Successful Partnering is the fine balance between 'non-contractual' working methods and essential formalities that ensures a good and productive working relationship with demonstrable benefits and all due diligence.

A Partnering contract tends to carry a higher amount of risk than a conventional contract for various reasons, some of which have been touched on above. The areas of higher risk with Partnering include, or could be seen to include:

- Tendency to be used in larger contracts.
- Working away from a formal contract arrangement.
- Contractors advising on spend patterns.
- Contractors sharing cost savings.
- Developing closer working relationships could lead to collusion.
- Partnering contracts are often more complex and involve more work and so there is a need to demonstrate (better) value for money.
- *Real* Partnering would dictate the main contractor/s partnering down the supply chain – this can be hard to achieve with multiple suppliers and can lead to greater complexity.

In addition, working across organisational boundaries brings complexities and ambiguities that can generate confusion and could weaken accountability. For example, there could arise:

- Conflicting organisational agendas which could mean...

- Uncertainty around roles and accountability leading to...
- Less Control, missed due to...
- Lack of reporting structures

Whilst the emphasis of Partnering is towards a more personal and mutual approach to contract management, the potentially higher level of risk (to all parties) means that greater demands are made upon the contract managers' skills and abilities.

Why could it go wrong?

Partnering will work when the people involved with it understand partnering; they must want it to work and will make it work. To ensure success the following key areas of potential weakness must be consciously avoided:

- Choosing the wrong people will ensure failure: trust and pro-activity towards problems will not develop – which are both vital factors for a successful partnering arrangement.
- On this basis, the organisation/s involved must be *ready* to partner. If the culture is not 'in place', putting it there has to be a prioritised part of the early preparatory work, through workshops, training and other general, attitude-changing exercises. The idea may have to be sold but, once sold, it has to be made to work.
- Objectives must be clear – an obvious analogy is that all members of a tug-of-war team must know in which direction to pull! The nature of a partnering contract means that all those involved have to (automatically) pull in the same direction without constantly being told.
- Clear methods of measurement have to be established at the outset so that the success or otherwise of the arrangement can be objectively judged. Where profit-sharing (and loss-sharing) in particular are part of the deal, an agreed and accurate measurement of performance against target or guaranteed maximum price (for example) is vital if disputes are to be avoided.

Help with setting up and running Partnering Contracts

Partnering, as a form of contract, is not an easy one to tender or manage and takes more time and effort on the part of everyone involved. However, when it works, the rewards are there to be had and the whole contract management process can actually be enjoyable!

The key to making Partnering work is communication and honesty. From honesty, trust will ensue. Each party has to recognise the needs of the other and work with them for their mutual benefit. It is harder work but can reap rewards conventional methods of contract management can only aspire to.

Advice and assistance should be sought from your legal section, your central procurement unit (if you have one) or your organisations 'expert' as well as, if possible, your Corporate Risk Management Section .

Summary
- Partnering is a method of service or project delivery where all parties to the contract share responsibility for its success:
- A basic concept is shared gains and shared responsibility for losses and other issues
- Often, Partnering contract tenders comprise a large proportion of quality elements
- Cost elements for projects normally involve prices submitted against a contract specification
- This price becomes the 'Guaranteed Maximum Price' against which savings are measured.
- Term contract costs are normally submitted against a Schedule of Rates
- The provider's profit and overhead percentages are known and often form part of the tender submission
- The process of contract management, cost control and share of profits depends on open-book accounting.
- Partnering is not easy – but if it works, it is worth it.

10 // Facing Challenges – The Standstill Period

What is a Standstill Period?

Known colloquially as the Alcatel period as a result of the precedent established in a legal case involving a company of that name and the Government of Austria, the Standstill Period was introduced to ensure that firms involved in an EU tender process were properly informed of the outcome and had the chance to query the result and get feedback on their own submission.

Whilst the requirement to observe a standstill period is only legally required for above-EU contract awards (and so, I have assumed, will require Cabinet or Member sign-off), it is good practice to follow the same procedure – a voluntary standstill period – for below-EU awards as well. It demonstrates transparency and gives bidders a chance to raise queries before you have gone too far. In such a case, as it is voluntary, you do not have to follow the Alcatel timescales to the letter.

The proper Standstill Period lasts for a minimum period of 10 days; it is a mixture of working and calendar days and this can confuse people, so we will look at this aspect carefully in a few minutes. Meanwhile, here is an overview of the Standstill Period.

Be warned – the Standstill Period seems simple enough, but the detail is complex. If you do not get the Standstill Period right, your contract can be cancelled or, as the Regulations put it, 'set aside'. Then you would be in serious trouble[1].

What do I do in the Standstill Period?

The Standstill period has five distinct stages, as defined by me in the timetable in Figure 10.1 below. These stages (and their timescales) are defined here on the basis of OGC Guidance issued at the time the Standstill requirement was introduced and we shall look at each of these component parts in turn.

---

1. See Ineffectiveness, Chapter 11.

Stage of Alcatel Process	Number of days
Start of Alcatel Process	Call it Day Zero
Unsuccessful bidders ask for debriefing and feedback and are able to raise queries	2 working days
Minimum period allowed for debriefing unsuccessful bidders	5 Calendar days
Minimum period between end of debriefing and the end of the Standstill period	3 working days
Standstill period ends – Contract Award can proceed.	

Figure 10.1 Alcatel Timetable

When does it start?

The Standstill period starts with the issue of 'Alcatel Letters' or Notices. These inform bidders of the outcome of the tender process, their performance within it and the intended appointee/s. When counting, the day the letters are issued is Day Zero.

When are Alcatel Letters Issued?

Technically, the letters cannot be sent until the decision to award has been made. In a Local Authority, whilst the report making the recommendation to award may be signed off (i.e. Approved), it will not normally be 'ratified', or the award will not become fact, until Executive, Member or Cabinet sign-off (depending on the value) and (when required) a Call-in or Scrutiny period (normally of five days duration). In theory – and in law – it is possible to issue Alcatel letters once the Award Report is signed off but there is always a risk: if the Approval decision is challenged and the decision overturned, you will have already declared information that should not be in the public domain (see later). It can therefore be advocated to *not* issue any Alcatel letters until the Call-in period is over and it is safe to do so[2].

2. Remember – Call-in periods only start after the decision is published in the minutes of the meeting. This in itself can take a week or more.

However, the likelihood of a Member not signing off a recommended award or Call-in changing a decision is normally remote: it depends almost entirely on the political landscape in your organisation so it is, to some extent, a judgement call on your part. The problem here is that Cabinet and Member decisions are published and become public information. They could 'spill the beans' before you have notified the bidders and that would be bad news. So, if you do wish to issue Alcatel letters before the decision is officially ratified you must make it clear in the Alcatel letters that the award is subject to ratification. The benefit of this is that the bidders know the outcome before it is made public.

This latter point is particularly salient when Section 20 consultation is required: you have to divulge details of the outcome in the Section 20 Notice of Proposal (see later) that really the bidders ought to know first. The law is silent on this chicken-or-the-egg conundrum as all the relevant pieces of legislation contradict each other, and everyone, like me, has a view and no-one is really right or wrong.

My advocation is to issue the Standstill letters to the bidders, with the clear caveat that the decision or outcome is subject to Cabinet (for example) sign-off and the result of Section 20 consultation. In reality, the chance of a challenge from the unsuccessful bidders is probably higher than from the other two areas of risk so putting the proposed outcome 'out there' with caveats is probably the best way forward.

What do the letters say?

I could provide a template for Alcatel letters, but your legal team ought to be able to provide you with a version that suits your own organisation's corporate style. The content of the letters, though, is quite specific in law.

To the winning bidder, you merely need to send them a letter advising them of their success and their 'score'. The 'score' will be their tendered sum if it is a lowest-price bid and their points score if it was a MEAT evaluation. You ought to give price and quality score separately.

To unsuccessful bidders, you need to send:

- The name of the winning tenderer(s)
- The score or price of the winning bid(s)
- Their own score or price
- Where their bid was weak or strong
- The offer of a chance to have further feedback on their submission
- Arrangements and contact details for this.

Be mindful – as I said above, if you *do* send out Alcatel letters before the end of a Call-in period, you will need to qualify your letter by advising the bidders that the decision to award is not final until the Call-in period is over without challenge. You should avoid inadvertently making the Alcatel letter a Letter of Intent[3] but may also wish to add that any expenditure the contractor may make in preparation for their new contract is entirely at their own risk. You can see how it all starts to get a bit messy.

To whom do the letters go?

Originally, letters had to be sent to all those involved in, or who 'expressed an interest' in, the contract, and this was an arduous task. However, a recent Amendment to the Public Contract Regulations[4] says that you no longer need to issue a standstill letter to any applicant or tenderer who has been excluded from the process, provided the limitation period for them to bring a claim has elapsed (i.e. a maximum of 3 months). To rely on and make most benefit from this amendment, you should ensure that any company who is so excluded (e.g. at the PQQ stage) has had it made clear to them in writing and at the earliest opportunity.

Should you be in a position to issue standstill letters before the three months limitation period has passed (and probably, therefore, in most instances when using the Open Procedure), you

---

3. Letter of Intent – see Chapter 14
4. The Public Procurement (Miscellaneous Amendments) Regulations 2011

will not be able to take advantage of this change. So, in detail, what does all this mean?

Sometimes you may get in excess of 100 firms 'expressing an interest' in a contract – i.e. asking for a PQQ. However, simply requesting a PQQ is *not* necessarily expressing an interest but merely seeking additional information and the *opportunity* to express an interest. It is generally accepted in practice that only firms who *complete and return* a Questionnaire have actually expressed an interest in tendering.

To keep everything in proportion, manageable and legal, accepted good practice is therefore to send Alcatel letters to (a) all those firms who tendered (of course) and (b) all those who submitted a completed Questionnaire but have not yet been advised of the outcome[5]. This can be a chunky process, but e-mail and – particularly – e-tendering systems have made this task much easier than it might otherwise have been.

The new Amendments can make this even easier:

Under the Restricted procedure

If using the Restricted Procedure, those firms who submitted a Questionnaire but will not be invited to tender (for any reason) should be promptly sent a letter (a 'Dear John' letter) advising them that they have been unsuccessful in their application and briefly explaining why. Feedback should be offered as good practice. Having done this:

- No Alcatel letter will be required if a 'Dear John' letter has been sent and a period of three months has elapsed before the Award decision is made (or rather – before Alcatel letters are issued[6]).
- You must make sure you send the 'Dear John' letters of advice and do not rely entirely on the three months' limitation period because...

5. OGC Guidance, January 2010
6. Be careful – deliberately delaying Alcatel letters to allow the limitation period to set in could itself be challenged as bad procurement practice or even collusion; additionally, it will delay the award of your contract.

A company can legally consider it has not been excluded from a process unless and until it has been so advised, and the courts will agree.

Reminder: If the Award decision or issue of Alcatel letters is made *less* than three months from the date of your 'Dear John' letters, then Alcatel letters must still be sent to all who expressed an interest.

Under the Open procedure

If you are using the Open procedure, *all* bidders who submit documents have technically registered an Expression of Interest (remember – the Questionnaire is included with the tender documents) and so all of these firms will require an Alcatel letter or notice of some sort.

Remember, in the Open Procedure, you only need assess the bids of those who pass the Questionnaire stage so, for those firms who do not pass the Questionnaire, you should promptly issue them with a 'Dear John' letter, just as in the Restricted Procedure, advising them that they have been unsuccessful at the Questionnaire or Qualification stage (as it is called in the 2015 Regulations), explaining why and offering feedback.

However, remember you may also choose to assess the bids first and then look at the Questionnaire. If you do it this way round, then you will issue Standstill letters that advise them of why their tender was not successful.

Two more points to note:

- If you do not reach the stage of Alcatel and Award within three months of this 'Dear John' letter of advice, then Alcatel letters do not need to be sent to these unsuccessful bidders: you will only need to send standstill letters to those who went on to have their actual tenders considered.
- If you *have* evaluated and prepared for award within 3 months of issuing the 'Dear John' letters (or you have simply not

issued them) then you will need to send Alcatel letters to all participants.

It is all down to timing – and your own efficiency.

General Considerations

Regardless of the procedure being used, Alcatel letters and 'Dear John' letters advising of failure at Questionnaire stage should – where possible – advise firms why they did not pass and / or offer the opportunity for detailed feedback. Make the content as detailed as possible: this will demonstrate openness and reduce the likelihood of requests for further feedback. However, it will not eliminate it...

It has been assumed that you will issue your Standstill letters electronically. If you issue them by normal post (why would you?) the Regulations say you have to allow 15 days – not the minimum 10 – for the standstill period. This – I hope – is of academic interest.

Feedback – What is it and how do I do it?

Feedback – sometimes called Debriefing - is essentially advising a firm why they were not successful. It is intended to serve two purposes:

a) To give them the opportunity to verify that the scoring process was good and fair – i.e. you got it right – and to raise queries about the outcome , and

b) To advise them and thereby assist them to prepare a better bid another time.

Feedback should be provided within 15 days of the request and can be given in a variety of ways: you can make suggestions and define a process but you will not have the final say – unsuccessful bidders may ask for something different and if it is reasonable you should assist. The tender documents will explain the Feedback process but the Alcatel letters need to explain it as well. Also, the depth of feedback people will require will be very different, and you will have to be prepared to answer some in-depth questions.

The easiest way to feed back is via e-mail, when you can compose a simple critique of a bidder's submission and send it off. Some firms are happy with this but you need to be careful because the feedback is in writing – do not write anything that can come back to haunt you. On this point, some people like to prepare a written critique of all non-successful bids, explaining how the scoring process arrived at the final outcome. This document is issued with the Alcatel letters or upon request and can save a lot of time at (my) third stage.

Some firms will accept feedback over the telephone, happy that their questions can be answered and concerns addressed quickly and simply.

Finally, feedback can be given at a face-to-face meeting and some firms will insist on this; they may ask for it from the outset or request it if they find the written or telephone feedback unsatisfactory. If the bidder turns up with solicitors or procurement experts – fear the worst!

Whichever method you employ you must never:

- Give them the name of any bidder other than the winner/s
- Give them the scores of any other bidders – just their own and the winner's
- Explain how the winner/s final score was made up

You can only discuss with them what the wining tenderer scored overall (points or pounds) and their own score and, in this latter regard, you can go into as much detail as you like.

Three tips:

1. If meeting bidders in person, avoid clashes in the waiting room: bidders should not encounter each other – that would be very bad form.

2. Ensure quality evaluators make clear and thorough notes when scoring bids – these notes will make your Alcatel letters and any additional feed-back much simpler to collate.

3. Avoid face-to-face feedback if you can.

Can I award the contract after 10 days?

Not really, no. I will give you a helpful tool to calculate how the actual days work out later on but in the meantime, looking at the timetable in Figure 10.1 above, you will see that the regulations lay down a minimum of three working days after the end of debriefing before the standstill period can end. It may be that you cannot see all the bidders within the stipulated five calendar days, and if this is the case the debriefing period will have to be extended. That is fine (apart from delaying the contract) – EU timescales are always *minimum* periods. The three days between the end of debrief and actual award will have to start when you *have* finished debriefing.

Challenges and How to Avoid Them

If a firm is not happy that you have got the process right, or if they feel that in any other way the evaluation was not fair or the outcome incorrect, they can level a challenge to the process and cause you no end of grief. This grief comes in many guises so beware those advocating the gung-ho approach that a challenge can "easily be defended". By then it is too late: if your procurement is challenged, you cannot proceed with the award process until the challenge has gone away. If legal proceedings are involved, you can guess how long that will be.

Challenges are levelled against a procurement process when it is seen to be flawed or when an aggrieved bidder thinks they can show it to be faulty (be they right or wrong), so you need to ensure that your procurement is demonstrably 'good'. This point is becoming more and more apposite as time goes by because:

- In a poor economic climate there is less work about so people will fight harder – and dirtier – for what work there is and they will probably have more time on their hands in which to do it.

- There is a current trend for society in general to be more litigious and construction and other companies are no different.
- The challenge culture is self-generating: as more challenges are seen to be successful so industry will pick up on the trend.
- The Remedies Directive of 2010 has now been incorporated into the 2015 Regulations but it gave courts in England hitherto undreamed-of powers, to the point where an upset bidder could have your contract nullified by an English court and not have to go (literally) to Europe at all. The Directive made these remedies – and so a challenge – much more accessible to aggrieved, unsuccessful companies.

Heed no other advice: the best mitigation against receiving challenges is to proactively avoid them and avoiding challenge begins at the very beginning of the procurement: the vast majority of (successful) challenges are founded on *process*. If you ensure from the very beginning that you have got the tender process right, you have reduced the risk of challenge by over 90%[7] . To this end, two key elements will mitigate the risk of challenge:

- A good strategy, including timetable, properly thought out and
- A good evaluation model, already subjected to challenge by your peers.

Chapter 11 will illustrate the high-risk areas of process but, as the saying goes, so long as you do it right and do not try to cut corners your process will be good. Always keep an eye on the procurement or purchasing press and see on what basis firms raise challenges: you will quickly get to understand the type of issues they seize upon to have an award decision overturned or have damages awarded in their favour. How one firm wins prompts

---

7. My own estimate, not officially or mathematically derived.

others to look in the same places for *their* chance to overturn an award decision.

Assuming your process is good, you need to be sure of your evaluation model so have your peers double-check it. Avoid ambiguities in the wording of evaluation criteria and leave no gaps. Make sure *any* aspect you are measuring has a score value identified and, if an evaluation has any area of subjectivity, make sure it is declared, along with the reasoning. Always double check your arithmetic.

Ensure all evaluation panel members understand the evaluation model and how it is to be applied – written guidance on this is always best and can be shown to bidders at feedback sessions (just make sure it does not say anything the bidders have not already been told, albeit explained from a different perspective). Ask evaluators to always enter scores against each criterion and to make clear notes explaining their decisions, particularly where it is perceived their judgement may later be called into question.

Always check the maths – too often I have found fault with the arithmetic in an evaluation model: if your model awards the contract wrongly through poor maths you really have no defence (at all, let alone in law!) and you may as well start packing straight away.

The rule of thumb is this: throughout the whole procurement, *always* be preparing for feedback: think 'how could I explain this in the face of a challenge or even, possibly, in court?' This way, you will always be making sure you can justify and defend, face to face, everything you have done. This is not a trick to avoid challenge – you really *do* have to be able to justify everything you have done.

You should try to make feedback meetings positive, amicable and pleasant; do not be defensive but be business-like and as brief as possible without being abrupt. Put simply, do not say more than is necessary. Avoid appearing as though you are hiding anything – simply avoid being sucked in to saying something you shouldn't that could be picked up, to your detriment, later on.

One final tool that can often help – especially if you think any feedback sessions could be 'sticky' (i.e. you sense a bidder is seriously aggrieved rather than simply disappointed) – is to have a pre-debriefing period meeting with the Evaluation Panel to look at the unsuccessful bids and agree the nature of the feedback. Apart from giving you confidence, the meeting may help you highlight key areas of weakness in a bid or it may provide you with useful phrases that can assist you with giving a bidder bad news more gently or more constructively than you might otherwise have done! All of this will help.

Making the Days Work

The timetable in Figure 10.1 above combines working and calendar days (Saturdays, Sundays and UK Bank Holidays are not working days) so, to make it easier to work out when that happens, I devised the table below (Figure 10.2).

To use this table, you will need a pencil. Fill in the left hand ('USE') column with the day and date with the pencil, making day 0 the day you send out the letters. You then count in all the days ahead, allowing for working and non-working days as necessary. To ensure you remember the difference, I have used red shading for working day periods and green for calendar day periods.

If anything changes, for example debriefing is delayed, Day 7 – end of debriefing – will move and all the days after that will change. *This* is why you fill it in using a pencil! You can rub out all the days after the change, enter the new ones and then know your new set of dates.

Once you become practised, you will not need such aids, but in the early days something like this may be welcomed because Alcatel is *not* simply 10 days, as some people keep on insisting!

Alcatel Calculator

Use	Day	Action
	0	Notify tenderers of Award Decision. Offer feedback. Advice letter to be sent by the 'quickest means possible', i.e. e-mail or fax. ALCATEL period must start (i.e. Day 1) on the day the letter will be received. If surface mail - Registered post is advised in such circumstances.
	1	Unsuccessful bidders* have **two working days** to request a debriefing session and raise queries.
	2	
	3	Debriefing period. Such debriefing is compulsory if the request is received within the first two working days of the Alcatel Letter. These are **Calendar** days.
	4	
	5	
	6	
	7	
	8	Three full working days are required from the end of the de-briefing sessions to the end of the Standstill period**
	9	
	10	
	11	Once the standstill requirements have been satisfied, contract may be awarded. 'Day 8' may be delayed and so therefore may 'Day 11'.
KEY:		Working Day Calendar Day

Figure 10.2 – Standstill Period Calculator

Summary

- The Standstill Period is often called the Alcatel period
- It comprises periods of working days and calendar days, so...
- Take great care to get the timescales of the Standstill period right
- Ensure you send Alcatel letters to all who tendered and 'expressed an interest' but in line with the requirements of the 2011 Amendments

- The letters have to contain, as a minimum, legally required information
- Make certain that the evaluation process provides sufficient information to assist the feedback process – scores, notes, etc.
- An evaluation Panel meeting may be of benefit to prepare for the feedback process
- Start avoiding challenges from the day you start planning the procurement

11 // Remedies and Ineffectiveness

I have stated quite categorically that this is not a book on law, and I have avoided any great excursions into legal clauses and jargon. It is, however, a book on how to work within the law and the Remedies Directive, introduced in 2009, had a major impact on the procurement process, hanging over the heads of procurement folk like the sword of Damocles.

When the EU Regs took on their 2015 persona, the remedies laid down in the 2009 Directive became part of the actual Regulations, headed 'Remedies', and so I shall cover them from that perspective.

The Regulations allow action under the Ineffectiveness Regulations to be instigated through the High Courts in the UK. The implications of this for a procuring client are obvious – any aggrieved bidder now has much easier access to legal remedies than they used to have. Although, of course, this is precisely why the changes were made, for us it means we have to sharpen up our game as a legal challenge is much easier to raise and so much more likely to occur than it has been in the past. Once an action has been taken against you, the (relatively) new Regulations mean that any legal finding against you could result in penalties much more severe than they might otherwise have been – and include the possible cancellation of your contract.

If you have the courage – read on.

What the Regulations Say
The Regulations invest surprisingly severe powers to the UK courts in the event of a breach of the EU Regulations. In brief, these Regulations state:

- A legal action under these new 'rights' must be started in the High Court.

- English Courts have much more regulatory power regarding procurement malpractice than it has hitherto had.
- These powers apply to all EU procurements commenced on or after the 20th December 2009. As the saying goes – that means you.
- There are different remedy regimes that apply before a contract is awarded and after a contract is awarded.

Remedies before Award of Contract

Before you have entered into a contract a Court will have the powers to:

- Set aside (i.e. cancel or nullify) any decision or action (e.g. the decision to award)
- Order the amendment of any document
- Award damages to any Economic Operator suffering loss or damage.

Additionally, such actions will be without prejudice to any other powers of Court – so consequential or additional claims may still arise from any other affected parties (e.g. bidders).

Remedies after Award of Contract

After you have entered into a contract the courts will have the power to:

- Issue a Declaration of Ineffectiveness (i.e. declare the contract void)
- Impose Civil Financial Penalties (i.e. a very large fine payable to the OGC[1])
- Award damages to bidders.

Be aware:

1.May now be payable to the Cabinet Office - see Chapter 16.

a) If a Declaration of Ineffectiveness is issued, a Civil Fine *has* to be imposed and the award of damages to bidders would be an almost automatic further consequence.
b) The most severe retribution – the Declaration of Ineffectiveness – is *mandatory* if any of four specific 'grounds of ineffectiveness' are proven. These grounds are:

1. Failure to publish a Contract Notice (OJEU advert) when required
2. A Contract is entered into in breach of Standstill (Alcatel) requirements
3. A Contract is entered into in breach of a Court Order or proceedings
4. A Contract is awarded under a Framework in breach of Regulations governing competition and price thresholds.

Note the word 'mandatory' in (b) above – the Courts have almost no choice *but* to take the actions listed above. The only mitigations for not imposing mandatory ineffectiveness are where:

- The ineffectiveness would be contrary to 'general interest' or the common good and / or
- Such action would have a disproportionate adverse impact, or...
- One or more of a list of caveats apply that can ameliorate the situation.

A plea for mitigation on the grounds that ineffectiveness would have a disproportionately adverse impact on the contract itself would *not* be considered suitable mitigation. However, and this is the good news, in practice, a contract will not be set aside immediately, but only after an appropriate period of time has lapsed to enable a new provision to be put in its place. Such consideration would take into account the 'common good'.

Get-out Clauses

Don't despair. As is always the case with issues in law, most of the four doom-laden grounds for ineffectiveness laid out above have caveats attached to them that may serve to help you out of a spot, and I shall look at these now. The law says:

- *Failure to publish a Contract Notice (OJEU advert) when required* is not a ground when an Authority:
  a) Considers a Notice was not required *and*
  b) Publishes a Voluntary Ex-Ante Transparency Notice (VTN or "Veet") *and*
  c) Does not enter into a contract for 10 days following the VTN.

- *Contract entered into in breach of Standstill (Alcatel) requirements* only applies as a ground if:
  a) The breach has deprived an operator of pre-contractual remedies *and*
  b) There has been a substantial separate breach of the Regulations *and*
  c) This has affected the operator's chances of winning.

- *A Contract is entered into in breach of a Court Order or proceedings,* by definition, means that the process was legally void from the outset and so there are no mitigations.

- *When awarded under a Framework in breach of Regulations governing competition and price thresholds,* grounds for ineffectiveness will apply unless:
  a) The Authority considers itself to be compliant (and they will have to prove in court that they do) *and*
  b) Undertakes a voluntary standstill period *and*
  c) Does not enter into a contract until the expiry of this standstill period.

It is important to note in the above lists of caveats that most of them are joined by 'ands' and not 'ors', and I repeat: Nearly all challenges to a tender outcome are based on errors or omissions in *process*, so use good process as your safeguard and do not rely on these get-out caveats as your protection. <u>*Do not*</u> get into the situation in the first place.

Voluntary Transparency Notices

Looking a bit like an OJEU Contract Notice, a Voluntary Ex-Ante Transparency notice (VTN) can be used retrospectively to advise the market of a procurement already undertaken when, by mistake, an OJEU advert was not placed. Aimed at avoiding an ineffectiveness suit, it does not necessarily protect you against claims for damages from other parties unhappy at the process you have undertaken.

VTNs are used – and more and more frequently – but there has been very little in the way of legal actions on which to base any advice on what the courts' views are. The best advice is to not need one: publish an OJEU Notice when you should and if you are not sure, publish one anyway.

Having breached the Regulations, a VTN is the legal solution. However, always bear in mind the fact that a VTN *might* cover or protect you, but on the other hand they do advertise the fact that you have breached and may tempt a challenge where one may not otherwise have lain. Looks like it's that judgment thing again.

Timescales

The Directive also lays down strict timescales within which an Action must be taken by an aggrieved bidder, and cases have been thrown out of court because the action was taken too late. However, this is still an area under legal discussion so keep an eye on new cases and note any fresh legal precedents that may arise.

The EU Regulations say that when the action is *not* seeking a declaration of ineffectiveness, an Action must be started promptly and, anyway, within 3 months of the date when the breach occurred.

However, Amendments were brought in for the UK[2] that require a challenge (proceedings) to be brought within 30 days of the date of knowledge. This 'date of knowledge' may be arguable but suffice it to say is retained as in the current understanding under EU Law[3]. Bear in mind, though, that the Courts still have discretion to allow claims to be brought in up to 3 months afterwards, if they see grounds for doing so.

The Regulations do not require proceedings to be started less than ten days after the company was notified of the decision or action in question and after it has been given the reasons for that decision, which links into and stretches the requirements of the Alcatel Standstill period.

In other words, even if you do not receive a challenge within the Standstill period, whilst you can go ahead and award, it does not mean that no challenge will be received or action taken against you. However, if no adverse approaches are received during Alcatel, it is less likely that a challenge will arise later on – unless, of course, it is an action regarding how the Standstill process itself was managed!

When the Action *is* seeking a declaration of ineffectiveness, an Action must be started:

- Within 30 days of an operator being advised of the award and being given the relevant reasons *or*
- Where the contract is already in effect, within 6 months of the day after the contract was entered into *or*
- Within 30 days of the publication of a Contract Award Notice (in OJEU) if – and only if – the contract was awarded without a Contract Notice (i.e. no OJEU advert) *and t*he Contract Award Notice includes justification for there being no OJEU advert. This generally applies when a contract is awarded directly (i.e. without competition) and will probably not be of great relevance to you at this time.

---

2. The Public Procurement (Miscellaneous Amendments) Regulations 2011
3. I cannot define this. Often 'reasonableness' is a point argued in Court. Sometimes, the date of knowledge is clear (e.g. an Alcatel letter) sometimes it is not.

Seeking Advice

Despite all the detail I have gone into here, *do not* embark upon any aspect of dealing with an ineffectiveness issue without the full support of your legal team. The information I have given above is – believe it or not – just a brief overview for your information: it does *not* qualify you as a lawyer in this aspect of procurement!

I repeat (again) – do not concentrate on how to get yourself *out* of a mess, just *don't* get yourself there in the first place. Do it *properly*.

Summary

- The Ineffectiveness Regulations give UK courts draconian powers in the event of certain breaches of the EU Regulations
- These include the possible setting aside or cancellation of your contract – after its award
- Additionally, Civil Penalties can be levied and damages awarded
- There are mitigations and caveats that might be deployed to avoid or defend some of the more serious court actions but...
- *Do not rely on these* – get the process right!

12 // Risk

Risk Assessment

The funny thing about procurement is that not many (ordinary) people know what it is yet just about everyone does it every day – most of us buy something and when we do we have choices, make decisions and comply with the law. I have bought many a theoretical pork pie when explaining complex procurement processes to people.

Risk is the same: the term 'Risk Management' has an aura of mystique about it but if we break it down to its component parts it is once again something we all do every day. Consider an example on a day you are going for an interview:

- As you leave in the morning, you ask (a common question): "Is it going to rain?" This is considering a risk – the risk of rain.
- The answer might be "probably not" or "maybe" or "possibly" or "probably will." This is considering the probability of the risk happening.
- If it *does* rain, what will happen? You may get damp, wet, soaked or, at worst, drowned. Here, you are looking at the impact of the rain, if it does come.
- So what can you do about it? You could stay in, take an umbrella, pack a Cagoule or wear a raincoat. The ways you might deal with the threat of rain are all mitigations.

Considering what risks there are and how they will impact is all part of the problem of Risk Assessment and this simple example above translates almost directly to even the most complex risk assessment processes used in procurement. Let us again consider the four key elements but in a slightly more serious manner:

- **Risk** is the chance of some event adverse to your goal (i.e. your project) happening.

- ***Probability*** is the likelihood of that event happening.
- ***Impact*** is the affect the event would have if it did occur.
- ***Mitigation*** is what you intend to do about it – i.e. managing the risk.

Impact is sometimes not entirely straightforward: it is certainly the effect the event would have if it occurred, but *what* effect on *what*? Consider the rain again: if it rains it could ruin your suit, make you look a mess before an important interview, slow down the traffic and generally make you look a poor candidate. These 'impacts', although from one cause (rain), are all very different: they have different effects and may require different mitigations. Consider them each in turn:

1) Ruin your suit: an issue of cost.
2) Make you look a mess before an important meeting: an issue of (your) quality.
3) Slow down the traffic: an issue of time (programme).
4) Make you look a poor candidate: an issue of your reputation.

You may therefore have to consider the impact of one risk on up to four different areas, and have a different mitigation or reparation for each one.

Assessing the Risk

It is obvious that, when there is a series of risks lined up against a project, the more severe ones will have to receive the most attention. To prioritise risks in this way, we need to allocate to each one a 'Risk Factor', based numerically on the combined effect of its probability and its impact. To do this, you need to allocate the probability and the impact a value from a predetermined set of levels. Unless you are using someone else's Risk Register (see later), *you* will have to set these levels.

Let us consider levels of Probability, how they are described (so that we all know what they mean) and what numerical value they are given. Look at Figure 12.1 below:

Likelihood or Probability	Level or Rating
Unlikely to happen, but could	1
Might happen	2
Likely to happen	3
Extremely likely to happen	4
Almost certain to happen.	5

Figure 12.1 – Level of Probability

I have included 'Likelihood' in the column heading as that helps people know what it means – the numeraphobes[1] amongst us would blanch at the very thought of the word 'probability'. We have a Rating range of 1 to 5 and no zero (which would mean no likelihood so it would not be there) and no six, i.e. certain to happen as this too would not be a *risk*.

We can look at Impact in the same way (Figure 12.2), again using Ratings valued 1 to 5 against common-usage descriptions:

Impact or Effect	Level or Rating
Minimal effect - e.g." a ripple"	1
Some effect but not serious	2
Will have a notable effect	3
Severe effect such as to knock the project off course	4
Devastating effect on the project	5

Figure 12.2 – Level of Impact

We arrive at the Risk Factor of each perceived risk by multiplying its Probability by its Impact to arrive at a number between 1 and

---

1. Numeraphobes: my word – people who hate numbers and the maths that goes with them.

25. The outcome of this is shown in Figure 12.3 below:

IMPACT	5	5	10	15	20	25
	4	4	8	12	16	20
	3	3	6	9	12	15
	2	2	4	6	8	10
	1	1	2	3	4	5
	Score	1	2	3	4	5
		PROBABILITY				

Figure 12.3 – Risk Rating Matrix

I have listed Impact ratings vertically and Probability ratings horizontally. Their product – the Risk Factor of any named possible event – is in the body of the matrix and these have been colour-coded to help with the process of prioritising them. The 'Risk Level' or 'Overall Risk Factor' or priority level of each risk is then defined as described in the table in Figure 12.4 below:

KEY	
	Severe Risk
	Serious threat
	Requires care
	Low risk

Figure 12.4 – Level of Risk or Overall Risk Factor

Risk Matrix

We now have all the components we need to set up our Risk Register or Risk Matrix. This Register lays out all the information we have accrued into one table as in Figure 12.5 below[2]:

PROJECT:			RISK REGISTER							
Last Update / Review:			Risk Manager:							

Date	Risk	Nature of impact	F	T	Q	R	Prob.	Imp	Risk	Mitigation	Cost £	Owner

Figure 12.5 – Risk Register

We shall look at the use of each column in turn:

Column Heading	What it means
Date	The date on which the Risk was first identified or entered on the register
Risk	Description of what the risk actually is, e.g. 'Site contamination'
Nature of impact	What will happen if the risk occurs, e.g. 'Works halted for 10 weeks'
F, T, Q, R	Identifies the risk impact as Financial, Time (Programme), Quality or Reputation-related. You should only deal with one area of impact per line. You would mark each different area of risk in a new row an assessment of (the same) probability and (possibly different) impact for each and its own mitigation.
Prob.	Probability Rating of 1 to 5.

2. Fig. 12.5 – Example only – in practice, the format would be landscape as the column widths would need to be wide enough to accommodate sufficient descriptive text.

Imp.	Impact Rating of 1 to 5
Risk	Risk Factor – Probability x Impact. This column has each cell coloured according to the Risk Factor, as shown in Figure 12.3 above. Software will enable you to set up these last three columns to calculate the Risk Factor and colour the cell accordingly. You monitor and deal with risks in accordance with the priorities indicated in this column. Different rows for impacts against F, T, Q and R will have different Risk Factors.
Mitigation	How you will deal with the risk – see below. Different impacts – again, F, T, Q and R – will each probably demand different mitigations.
Cost	How much the risk will cost, including the proposed mitigation and the element of risk remaining- the *residual* risk (see later). This could actually be the cost of the risk itself - always use maximum figures
Owner	Every risk must have an owner, under whose remit lies the monitoring and management of that risk.
Last Updated	Last date the Risk Register was reviewed
Risk Manager	Person responsible for maintaining the Risk Register (*not* always a risk owner)

Figure 12.6 – The Meaning of Column Headings in a Risk Register – from Fig. 12.5

Mitigating Risks

There are various ways to reduce or minimise a risk and sometimes there are ways to remove them altogether. These 'ways' are called

Mitigations and these can reduce the likelihood or the impact.

Sometimes a risk can be mitigated by transferring it directly into the hands of the contractor or an insurance company; this can work but remember this will *always* cost you money. If you retain the risk you may *not* incur the expense (after all, it is only a risk). In general, mitigationscan include:

- Simply accepting the risk – take no action (not often recommended).
- Reducing or Minimising the risk – by having back-up and contingency actions.
- Getting rid of the risk completely – through a major change of plan.
- Transferring the risk to the contractor.
- Insuring against the risk.

There is no rule about which approach you take: you need to analyse the options and make a reasoned best decision at the time. You can always change the proposed course of action later.

Managing Risks – The Risk Register

How it operates

The Risk Register should be started or 'opened' at the very beginning of the procurement process – at the same time as and feeding into the Procurement Strategy[3]. Whilst a range of people may be risk 'owners', the Risk Register is always 'controlled' or managed by one, named individual who should be identified on the Register.

The Register should be reviewed regularly – certainly at every Project Meeting: new risks should be identified, logged and assessed, existing risks reviewed and updated. The latest update time should also be identified on the Register.

---

3. A Procurement Strategy should take account of risks and serve to deal with or mitigate them.

The risks you must consider should not just be the (more obvious) threats to a successful procurement. You must also consider risks that might arise during the contract period itself – in many cases these risks can be mitigated by actions taken (mitigations) at time of procurement. An obvious example of this is the risk of the contractor going into liquidation: you can mitigate against this by ensuring you seek a Bond or Parent Company Guarantee[4] from the provider as a condition of contract award. This will not prevent the incident occurring but will safeguard the client (a form of insurance) against the costs and some of the service-delivery implications that would arise from it.

Formats and Definitions

To explain the principles I have used a fairly simple Risk Assessment model and it is perfectly workable and fit for most purposes. There are, however, more ways than this to skin the Risk Register cat and the model can almost be what you want to make it. I will look at some alternative versions for you so that you will [a] not be phased when you see different or more complex versions and [b] be able to consider and develop different models for yourself in the future.

My examples above use a model based on ratings ranging from 1 to 5, but another common format uses a 1 to 4 range, like this:

Percentage Chance	Probability	Impact	Rating
1 – 10%	Remote	Minor	1
11 – 49%	Possible	Moderate	2
50 – 89%	Probable	Significant	3
90 - 100%	Certain	Severe	4

Figure 12.7 – Alternative Risk Factor Levels

---

4. See Chapter 16

You will see here that the definitions are also different to mine, but they still have the benefit of 'common use' language – a practical approach to assessing the risk. I have also shown how likelihood (probability) can also be expressed as a percentage chance of something happening – another option that may be suitable in certain circumstances. When using percentages, it is normal to specify the timeframe within which the chance of an event occurring is based (e.g. within one year, before start on site, etc.) to give the values some solid context.

Note, however, that despite the quoted percentage chances of an event occurring, a 0% chance means it will not happen (no chance), so it is no risk at all and 100% chance means it definitely will happen – this again is not a risk but a fact and so needs to be dealt with as such.

The Ratings of 1 to 4 will give a maximum Risk Factor of 16, so our Register will need to asses the resulting Risk Factors on that basis, as in Figure 12.8:

Overall Risk		
Value	**Term**	*Description*
1 to 6	**Minor**	Adverse but manageable affect on programme
7 to 12	**Major**	Warrants consideration & action but will not cause programme to fail
13 to 16	**Critical**	Risks programme failure

Figure 12.8 – Overall Risk Definition, Alternative Model

In this Figure (12.8) I have also taken the opportunity to demonstrate a different way of defining high and low risks[5].

---

5. Although included for numerical completeness, a Risk Factor of 7 cannot, of course, be achieved. I hope you can understand why.

More Complex Formats

All of the Risk Analysis models so far have been fairly simple and perfectly adequate for most day-to-day applications. However, if the application is more critical – more risky, more expensive, higher profile, perhaps – then the model may need to get more complex.

Simply identifying a mitigation does not always remove a risk: it may reduce it but still leave what is termed a *residual* risk and a more complex Risk Register will take account of these. To do this, you simply take the simple model I have given above and add, probably after the 'Mitigation' column, three more columns that re-identify the Probability, Impact and so the Residual Risk Factor once the initial mitigation has been enacted. An example is shown in Figure 12.9 below.

PROJECT:							RISK REGISTER						
Last Update / Review:					Risk Manager:								
Date	Risk	Nature of impact	F	T	Q	R	Probability	Impact	Risk	Mitigation	Probability	Impact	Residual Risk	Cost £	Owner

Figure 12.9 – Provision for Residual Risk in a Risk Register

The overall Risk Factor should reduce, otherwise the mitigation is not effective but you will find that mitigation will often (not always!) only reduce the likelihood of an event happening rather than reduce its impact: this is logical - if it still happens, the impact will often be the same. On this basis, when assessing risk using such a model, be 'smart' about what effect the mitigation has on the risk.

Figure 12.10 gives you another example of a 'complex' model, giving you a good insight into how you can be quite specific about impact ratings in specific areas rather than generic as we have been above. You will see that the model is again based on a 1 to 4 rating and approaches the question of impact – and describing it – in yet another different way. Compare this with Figure 12.8, for example.

Impact Rating	Failure to provide statutory duties or meet legal obligations	Financial loss	Service Disruption (Days)	Personal privacy/data protection infringement	Reputational Harm	Impact to Service User/s
ASSESSMENT OF IMPACT						
4	Litigation/claims/fines: Departmental £250k + Corporate £500k +	71 - 100% of Annual Budget	Over 5	All personal information compromised/revealed	Officer(s) and/or Members forced to resign	Death, or major permanent incapacity or disability
3	Litigation/claims/fines from Departmental £50k to £125k Corporate £100k to £250k	31 - 70% of Annual Budget	4 to 5	Personal Information of many individuals compromised/revealed	National public or press aware	Serious injury or long term incapacity or disability
2	Litigation/claims/fines from Departmental £25k to £50k Corporate £50k to £100k	11 - 30% of Annual Budget	2 to 3	Some individual personal information compromised/revealed	Local public or press interest	Significant injury or ill health requiring medical intervention
1	Litigation/claims/fines from Departmental £12k to £25k Corporate £25k to £50k	0 - 10% of Annual Budget	1	Isolated individual personal information compromised/revealed	Contained within section or department or division	Minor injury or ill health requiring first aid or self treatment but no incapacity

Figure 12.10 – Another Model: Impact Ratings

Still keeping with the 1 to 4 rating levels, Figure 12.11 below now takes a different approach to defining Risk Factors, using the level of response required as opposed to saying how serious an event would be, and this can work well although it is generally less flexible than other models might be. For example, by definition, a 'High' risk would have to be reduced to 'Medium' for it to be 'parked' and monitored, rather than dealt with immediately, which may sometimes be the desired approach even with a very serious risk. Of course, you could defer your 'immediate action', but then what is the point of having the definition?

ACTION TO BE TAKEN AFTER SCORING		
Risk Factor	Rating	Action
1 – 4	LOW	Limited action, incorporate into long term plans
5 – 8	MEDIUM	Review previous controls/actions – incorporate into current plans
9 – 16	HIGH	Immediate action required

Figure 12.11 – Another definition of Risk Factors

You will by now have realised that there are as many Risk Register models as you may wish to create and you can complicate to your heart's content. My own view is always that simplest is best provided it is adequate for the job and this matter of 'fit for purpose' is for you to decide. Avoid making it over-complicated just because you feel you need to create some impression of managing risk better. The Register is a just a log – it is *you* who makes sure the risk is managed.

One Obvious Point

I have shown you examples to illustrate the nature of the detail required in a Risk Register. When setting out one of your own, it will need to be in landscape format and have enough column width to contain all the required detail of risk, impact, mitigation, etc. This is obvious, but I thought I should say it.

Risk Management

You can see from this that Risk Assessment, from which you draw up and maintain your Risk Register, can be as simple or as complex as you choose to make it, just so long as it is suitable for the task in hand. This Assessment process is all part of managing a risk: the period when you identify it, log it and assess it. You then have to consider mitigations to remove, reduce or cover the risk and then enact them. As project leader, you will have responsibility for ensuring that risk is managed properly, that all risks have an owner and that the Register is reviewed on a regular and frequent basis: you do not necessarily have to be the risk manager or a risk owner.

You should aim to have one initial meeting of all (relevant) stakeholders. 'Relevant' can be a provocative word so I use it carefully – miss out one wrong person and all hell could break loose. By relevant I mean those with sufficient experience and knowledge to understand the meaning of risk, what problems might befall your procurement and so what risks need to be considered.

No risk should be discounted – if it is there, it is there – and the group should agree on the ratings and eventual Risk Factor. After that, individuals can identify new risks as they become apparent and the project group should as a team review the Register at Project Team meetings as a standard agenda item.

If a mitigation requires money, as project leader it is your job to ensure that the funds are available. If they are not, that proposal ceases to be a viable mitigation and another one will have to be found.

Do all this properly and the chances of your procurement going pear-shaped or suffering a challenge will be very much reduced.

Examples of Risks
It is easy to talk about risks, but some people find it hard to identify them – at least until somebody gets it started. To that end I have

listed here (Figure 12.12), in no particular order, some (generic) risks that might impact on any procurement or the contract that comes out of it. All of the risks I have listed here can and should be mitigated as part of the procurement strategy.

Possible Risk	Potential Impacts
At Procurement Stage:	
Lack of market interest Lack of providers	Insufficient tenders High tender prices
Tender programme runs late/gets delayed	Contract starts late Legal obligations not met Non-compliant process used to catch up Reputational damage
Tender sums higher than anticipated	Budget shortfall Service standard/specification threatened
Shortage of raw materials or Delays in delivery of materials	Impact on project programme Impact on budget
Objections to proposals (e.g. at planning or consultation stages)	Delays and threats to project Design changes incur additional cost
Services Part A contract wrongly identified	OJEU Notice not issued when it ought to have been so... Contravention of EU Regulations Tender or Award subject to Remedies Directive
Non-EU Tenders come in above EU threshold	Contravention of EU Regulations Tender or Award subject to Remedies Directive
Challenge during Standstill Period	Delays to the award process Change of tender outcome Delays to start of contract Penalties and damages incurred
Insufficient time for full legal procurement process	Contract starts late Legal obligations not met Corners cut Vulnerable people at risk

Low-price bid results in lowest quality bidder succeeding	Quality of service suffers Demand on client Contract Management increases High level of customer complaints arise
At contract stage	
Provider goes out of business	No service provision Project halts Costs incurred to tender and re-instate the provision
Organisation's needs change	Contract no longer meets requirements
Lack of client contract management skills	Provider does not meet obligations Poor quality of service
Provider subject to takeover or buy-out	Need for Novation New provider does not meet corporate standards Possible legal need for re-tender
Late challenge on the tender process	May be subject to the Remedies Directive Penalties and damages incurred
Contractor fails to perform	Standard of service provision too low Vulnerable people under threat Client falls liable for service failures
Joint client's increasing demands cause service in your area to suffer	Reduction in your service quality Provider begins over-trading

Figure 12.12 Examples of Risks

Notes on the impacts in Figure 12.12

Whilst I have shown some possible impacts that the risks could have, the impact in reality will depend on the nature of the contract you are tendering. I have not therefore suggested any mitigations nor have I suggested any probability or impact ratings – you will need to formulate these in the context of the service you are tendering and your own organisation. However, additional comments on some of the impacts I have described above may be useful.

a. If non-EU tender values come in above the EU threshold, there is a potential contravention of EU Regulations. As explained elsewhere, if the cost is *marginally* above the EU threshold and it can be demonstrated that good process has been followed and the error was unintentional, it may be in order for you to proceed with award. A rule of thumb says 'marginally' means about 5% but legal advice should always be sought in these instances and latest legal precedents may steer the final decision. If it is decided you *are* in contravention, the tender will have to be run again as an EU-compliant process or a VTN will need to be issued – and that has its own risks attached (see previous chapter).

b. If a procurement runs late, the temptation may be to cut corners. This may mean shortening tender times, speeding up evaluations, and so on. This is natural but be aware that any shortcut involves an element of risk in its own right – in fact, taking short cuts ought to be in the list above, really. So short cut at your peril, although I know it might be unavoidable if an unforeseen delay occurs. A good timetable should lay down the times you consider a process should take and include contingency times to allow for delays. *Starting* a procurement too late is another matter altogether, and so not an issue for discussion here.

c. It can be seen, even in this list of simple examples, that the *scale* of impacts can vary widely from possibly relatively minor 'delay to award process' to more serious 'failure to meet legal obligations' to the frightening 'Tender or Award subject to Ineffectiveness'. Reflect these differences in the rating but do not leave out a risk because it is 'less important'. If it happens it *will* be important.

d. Prevention is always better than cure (how often have I said this?). However, where there are legal cures for a risk, make sure the Register reflects this. For example, not issuing a Contract Notice for an above-threshold procurement immediately puts you under threat of measures under the Ineffectiveness

Regulations. The first mitigation might be to check again whether it requires a Notice (prevention). If you check and then tender without one, the risk is *still* there: the likelihood is *much* less (we would hope) but the impact is *much* greater and the mitigation would be have to be the retrospective, legal options I have described in Chapter Eleven. This illustrates the need to update your Risk Register on a regular basis – things do change.

Summary

- A risk is an event that might occur which would have an adverse impact or effect on your project.
- Risk Management is part of and, more importantly, helps steer your procurement strategy.
- You also need to consider contract risks at time of procurement and try to 'design them out'.
- The product of Impact and Probability will give you the Risk Factor – the higher the factor the greater the risk and the greater the need for mitigating action.
- Risk Registers can be simple or complex – make yours what it needs to be.
- Risk Management is not an additional task; it is an inherent part of good procurement practice.

13 // Sustainability

What is Sustainability?

Once upon a time it was a minor consideration, now sustainability and the reduction of carbon emissions (an organisation's 'carbon footprint') is one of *the* key factors in any procurement.

News broadcasts constantly report on environmental matters and climatic change and these really do constitute what may be called the 'big' sustainability issue – the sustainability of our planet earth. But for us sustainability is – paradoxically – more wide-reaching than this and has much wider implications. We have to consider, without ignoring the planet, sustainability issues on a more immediate and local level and in more areas than just climate change.

The subject of sustainability cannot be fully covered in one chapter – nor even in a whole book. It is vast and you are advised to make use of a copious number of books, guidance documents, seminars and so on to get a grip on the (much) bigger picture. But what I can do in this chapter is give you enough information to put you on the right road – to equip you to understand what sustainability is all about, how it impacts upon procurement and how to set about dealing with it in your tenders.

Sustainability for us has three key aspects for us:

- Environmental
- Economic
- Social

We shall look at each of these in turn.

Environmental Sustainability
I opened this chapter with a reference to climate change and, when tendering, due regard must be given to the environmental impact of the service we are specifying. Of course, the intensity of this impact will depend on the nature of the contract but construction (with which we are mainly concerned) is one of those areas where the

environmental impact can be most severe but it is also one where the most effective measures can be taken to minimise these effects.

Service Contracts

In construction and major repairs work, noteworthy measures can be taken to reduce carbon emissions and minimise environmental impact, and we shall consider these in some depth later on. In your position, however, you may also be called upon to tender Service[1] contracts which, under EU legislation, include day-to-day repairs contracts[2]. You will find that things can be done to support corporate environmental policies – many things – but they may not all be as 'dramatic' or headline-grabbing as your efforts on a building site might be.

We shall be looking primarily at construction and major works projects, but you will see that a lot of the ideas we discuss can be picked up in service contracts. Obvious ones include the use of eco-friendly vehicles, the division of the service delivery into logical, geographic areas and the use of sustainable materials wherever possible, but you will need to translate what we look at for construction work into a format suitable for service contracts, where this applies to your procurement.

Service contracts also include the provision of consultancy services and the measures you can seek to reduce environmental impacts when commissioning these services will be even more limited – reduction of office waste is an obvious one and paying due regard to reducing the environmental impact of contracts under their care is another. Generally, you may have to be satisfied with some general requirement in line with your own policies and leave it at that.

Environmental Considerations in Design and Specification

We shall look at two distinct aspects of construction: the design and specification stage and the on-site construction process itself.

---

1. Capital 'S'
2. Under EU definitions – the provision of a responsive repairs service, although it can look and feel like works, is just that – a Service, and it must be tendered as such

I realise that design will probably be outside of your remit, but there are certain issues you need to be aware of in the interests of carrying out a 'good' procurement and it may even be up to you to ensure that those who do design and specify are fully aware of all the issues that pertain and what options are available to them. You will find that, very often, the responsibility for making people aware of matters like this will fall into the hands of the procurement team – I'm never sure why.

Suffice it to say that energy efficiency, access to public transport, re-use of rainwater from the roof and so on are all environmental issues that can be dealt with at the design and specification stage. Of course, nothing comes free and it is a fact that, if you make a building more energy-efficient (for example through improved insulation) then the cost of construction will go up. The argument is that running costs will go down and the environment will benefit (depending on how the insulation is made, of course, and the resulting carbon footprint of the product. Complicated, isn't it?).

The best compromise between all the relevant factors has to be identified and this argument holds true across all aspects of the design and specification process: recycled materials or those supplied from a renewable source should be used but they are more expensive than those sourced less responsibly. Service contractors could use electric or gas-powered vehicles but will you have to pay a higher tender sum for the fleet? If you build near a good public transport hub, car-use will be reduced – but the building land will be dearer and construction costs higher[3]. Can you afford your environmental wish-list and where do you strike the balance?

Many decisions like these will have to be made and the calculations can be complicated: many of these decisions are now made on the back of an exercise called Whole Life Costing (WLC), and we shall look at this now in more detail.

---

3. Busy commercial centres make site access more difficult.

Whole Life Costing

The OGC[4] defines Whole Life Costing[5] (WLC) as the costs of acquiring a facility (including all consultancy, design, procurement, construction and equipment costs), the costs of operating and maintaining it over its life and the cost of the building's disposal when its life is over. WLC considers the cost of internal resources and departmental overheads, risk mitigations, flexibility (e.g. predicted alterations for known changes in business requirements), refurbishment costs and the costs relating to sustainability and matters of health and safety. In other words, WLC considers total ownership costs over the whole life of the facility.

Why do we have it now and not before?

WLC has always been here but, as already explained, the key drivers in procurement are changing and these changes are bringing sustainability and Whole Life Costing to the fore.

The pressure to achieve greater efficiencies in Public Sector procurement[6] has led to changes in the way the costs of a construction project are calculated, and the erstwhile tendency to isolate initial capital construction costs from the ongoing revenue costs of maintenance and operation is being reversed.

The drive for greater sustainability in all its guises has also led specifiers and designers to consider better usage of materials during a product's build and operation and more eco-friendly disposal methods once its useful life is over.

Whole life costing in construction

Whereas once upon a time capital expenditure was contained in order to maximise the number of projects one could fund from a budget, it is now recognised that additional expenditure at the design and construction stage can reduce the future, and thereby

4. The Office of Government Commerce – leads on good practice in public procurement. See Chapter 16 on the changed status of the OGC.
5. Whole-life Costing and Cost Management – Achieving Excellence in Construction, Guide AE7 (2007)e in Construction Procurement
6. Gershon Report 2004; Roots Review, 2009

total (i.e. actual), costs of the facility by significant amounts. A prime example of this, already alluded to, is the increased capital cost of improved thermal insulation in a building's roof, walls and glass; however, once installed, the ongoing energy costs associated with heating the space are reduced and the facility's potential residual (i.e. re-sale) value increased. By reducing the heat-loss we are, in direct proportion, reducing the energy requirements to heat it. In addition to any (revenue) cost savings and capital-value enhancement, the positive impact on addressing issues of carbon emissions and climate change is considerable.

The life-end of the product (i.e. the building) is also key to WLC considerations – i.e. how it is to be disposed of at the end of its useful life and what will the impact of that be? The products that go into the construction – under an effective WLC strategy – will be recyclable to the point where, again, both the disposal costs and the environmental impact are minimised.

Designing for minimised Whole Life Costs

The reduction of WLC begins at the very earliest stage of design. In general terms, WLC can be minimised through:

- Building in versatility so that alterations to the building fabric are minimised in the event of a change of use or occupier.
- Creating structures using materials that minimise maintenance and repairs costs, e.g. self-cleaning surfaces, erosion-proof components, etc.
- Achieving a specification that minimises energy costs for heating, lighting, etc.
- Using materials that, once the useful life of the building has ended, can be re-used or efficiently re-processed at minimum financial and carbon cost.
- The use of local (or localised) resources to minimise travel, labour and materials delivery costs (this during construction and the use stages).
- And so on.

How do we work out the costs?

The calculations can be as simple or as complex as the circumstance demands or you make it and they are dependent upon the skill of the person undertaking the exercise. In principle, all costs associated with the lifetime of the building need to be considered, although of course this will involve a good deal of speculation when future costs are being evaluated.

A list of the items that typically could or should be considered when working out WLCs would include:

Construction Phase	Use phase	End-of-use phase
Design costs Construction costs – labour Construction costs - materials Project management fees Temporary works Finance costs Fit-out and furnishing*	Cost of FM Team Utilities – gas, electric, water Insurances Responsive repairs Planned maintenance Cleaning Alterations Grounds maintenance	Sale profit or loss Demolition costs Change of use costs Site clearance cost Recycle / re-use costs or income Arisings disposal costs

Figure 13.1 – Whole Life Costing Considerations

* It would be a matter of policy and circumstance whether to lay Furnishings and fittings (F&F) costs at [a] the door of building provision or [b] within the operating costs of the business itself. If you were a developer letting or selling an empty building, it would almost certainly be the latter. If you were an organisation constructing the building and then occupying it, you would actually have the choice of including F&F in your WLC calculation or deciding that the business itself should fit out or equip the building from its own turnover as an operational expense.

Care needs to be taken to avoid including the costs of the *business* with the costs associated specifically with the building or facility. For example, the staff costs for an organisation are a business cost, not a premises cost, as would be telephone bills, etc.

A proper WLC calculation will involve an element of options appraisal: an incremental increase in capital expenditure would reduce the anticipated revenue costs of, say, heating but only up to a point where further tangible cash (or environmental) benefits fail to be realised. A decision on how far this capital investment should go would be made based on the following considerations, amongst others:

- How accurate are the energy-cost forecasts?
- Can the additional capital be afforded?
- What is the scale of the revenue savings?
- What is the company policy on sustainability?
- Is the organisation going to use the building long enough to recoup the costs?
- Do the enhancements increase the net worth of the building?
- What is the environmental 'cost' of manufacturing the insulation?

In addition, the WLCs should include an element of risk provision, calculated as part of a formal risk assessment and preferably on an element-by-element basis.

Whole life costing in a non-construction environment
Whilst the context is different, the principles described above can largely be applied to supplies and term contract procurements as well. In supplies contracts, the similarity is striking: the 'design' (i.e. specification) stage must include consideration of a product's lifespan and its packaging, operating, maintenance and disposal costs. Can the product be disposed-of in a manner that is ecologically sustainable? If so, is this going to be a more expensive option than simply making it from materials that can only go to landfill? What will the budget allow?

Term contracts, whilst less comparable, may primarily look at the WLC of the products or materials it specifies. With regards to the ongoing costs of the service provision itself, dividing a contract into geographical packages can reduce travel costs for, say, a responsive maintenance contract whilst longer contract terms will help reduce the overall impact of set-up and re-tendering costs, and so on.

Working it out

When asked to carry out a WLC exercise – or when you opt to do so – as part of a procurement, if it is to have any value at all it will need to be part of the early, project-planning process, feeding into any options appraisal and budget projections. How influential the process is in the final strategy decision will depend mainly on the priorities of the organisation in which you work.

The calculations can be complicated. Using spreadsheet software or employing a proprietary, purpose-built package, you can feed different values for each element of a contract into the model to assess the different permutations of current or immediate capital costs against future revenue costs for each element. The model will then give you the adjusted outcomes.

As already indicated, a lot of the process involves judgement and – yes – guesswork about future costs. This is fine so long as you make it clear on what basis you have made your judgements so that, as time goes on, known variations from your assumptions can be fed into the model and revised figures obtained.

Not yet an exact science, WLC is heading that way as practitioners become more adept at the process and more relevant data becomes available. As I so often say – you do not need to be the expert because you can seek the help and advice of someone who has the required levels of skill and knowledge in this area of expertise.

Is WLC the same thing as Whole Life Value?

No. Whole Life Value is more directly linked to issues of service, benefit in kind and sustainability, and is not part of a financial analysis (although costs, of course, are always relevant to what can be achieved in practice).

Of interest, though, is the consideration of Whole Life Value (WLV) in the context of tender evaluation, where quality and even that elusive holy grail 'added value' are sought, and considerations of WLV against WLC becomes an interesting reality[7].

The process of WLV assessment is still in its infancy but in principle assesses the value of the product or facility over its lifetime to *those it serves*, in terms of social, economic and environmental well-being.

This concept can best be explained through an example. Consider the fabric of a school building and how it relates to the Government's 'Every Child Matters' initiative[8], which advocates that every child should:

- Be Healthy
- Enjoy & Achieve
- Stay Safe
- Achieve Economic Wellbeing
- Make a Positive Contribution

The element "Be Healthy" relates (for example) to space requirements and comfort, and comfort requirements break down into light, heat, ventilation, cleanliness and colour.

Whole Life Value (WLV) will relate these elements of *building* design and performance to eventual pupil achievement (e.g. "effective heating and ventilation improves cognitive function by preventing heat loss or overheating").

WLV will identify the long-term *value* outcomes of the building and input these into an early stage of the design process of the project.

Back to Procurement

It can all seem a little overwhelming to the novice but WLC analysis is getting higher on the procurement agenda. It is too large a subject

---

7. The reader is referred to chapters covering MEAT evaluation criteria.
8. See: http://www.everychildmatters.gov.uk/aims/

to be considered in any greater depth here, but you do need to be aware of it, where it fits in and how to deal with it.

Learning More

You may or may not carry out a WLC exercise for a construction procurement, but knowing the issues it would raise will give you a good steer on where construction materials specification on your project should be heading. There are various organisations devoted to minimising the use of unsustainable materials and reducing carbon emissions and these are listed in Appendix 5. I recommended you research them – many of them provide practical assistance at little or no cost and will be only too happy to provide seminars and training to people in your organisation.

Some of these organisations also have an interest in the building stage of a construction project, where environmental issues can be tackled head-on, and we shall look at that aspect of environmental sustainability now.

The construction process

Key areas of sustainability that can be tackled within the tender and contract documentation include:

- Minimising waste on site.
- Sorting and separating waste into types or categories for recycling thereby...
- Minimising landfill.
- Minimising lorry movements through 'smart' materials ordering, including
- Buying from local suppliers wherever possible.
- Using local labour to save on journey miles.
- Controlling noise and dust.
- Controlling mud on the road.
- Emphasising site safety.
- Re-using demolition materials ('arisings') on site.
- Reducing use of – or reducing waste of – water and electricity on site.

Of course, many of these ideas actually cost money and so a 'policy' will have to be adopted as part of the procurement strategy that balances these costs and the benefits, just as we have discussed above, and your contract manager will have to ensure that your requirements are met on site. It is for this reason that, wherever possible, the contract manager is involved in the procurement process (see Chapter Two) – they need to have input and know what has been specified.

Achieving the result
There are a lot of things you can do in practice and procurement needs to consider all of them. Some of the issues may come up in the Risk Analysis (See Chapter 12), some may be matters of your organisation's corporate policies and some may be issues relevant to this particular project. Whichever applies, there are various ways to deal with environmental sustainability in your tender.

1.   At PQQ (or bidder selection) stage

At PQQ (or bidder selection) stage you can set parameters for the way the firm operates: you could insist on ISO14001 accreditation (an international standard for a company's environmental management system). This is a high standard and would only normally be demanded if the contract was particularly environmentally sensitive – a dirty homes cleaning contract or an asbestos removal project for example. To demand it for a standard construction project may be considered excessive although many of the larger firms will have it.

If ISO 14001 is excessive, you may consider stipulating a corporate environmental policy that matches at least your own organisation's minimum requirements (you would have to say what they are). You can develop this idea and insist on additional considerations within their policy that are specific to your project.

If the contract is not of high value and smaller firms will be bidding, setting your environmental bar too high may prevent good firms from passing through to tender stage: always keep things in proportion and if it is appropriate, simply ask to see the company's environmental policy as it stands. Explain that it has to meet what you consider to be the minimum standards an environmentally conscientious company would adopt and explain what these standards are.

In reality, for (really) small firms, having a policy at all would be a good start; sometimes they will have nothing at all (not a good start) and often they will only have an Environmental or Sustainability Statement. This is basically a statement of good intent but it may still be good enough for your needs. In reality, it may be that a bidder will draw up a policy specifically to meet your requirements and that is fine – "aim accomplished" so long as they are seen to work to it at contract operation stage.

With consultancies, you may find most of their environmental policies will relate to minimising office waste, which is natural, but you will also want to see that they recognise they must strive for wider and more beneficial environmental objectives in the work they do for you and your project.

2.   At Tender Stage

At Tender Stage you have two approaches open to you and you can adopt either or a bit of both. At PQQ stage you will have decided that their company approach is suitable, now you want to see what they propose to do for your specific project once on site.

The first approach is the obvious and easiest one, where you tell them what they will do within the project specifications, the preliminaries and the contractual terms and conditions. This is fine, relatively easy to do and turns the whole matter into one of contract management. It will work fine provided you have done it right.

The second approach is to move the subject into the tender evaluation model and ask bidders to submit proposals and method statements explaining how they will deal with specified sustainability issues on site. You should see Chapters 5 and 7 for more information on how to set the questions and carry out the evaluations, but I will say here that you need to be quite clear on what sustainability issues you want their method statements to address and what you will be looking for when you evaluate the submissions.

The more environmentally sensitive the project or the site, the more value or weighting the sustainability element of an evaluation will have.

I have stated before that matters addressed at PQQ stage cannot be revisited at tender stage and they can't. You must make it quite clear (in the way you phrase things) that the tender evaluation questions constitute a quite different issue to the matter of a bidder's corporate stand on sustainability in general.

Economic Sustainability

I have dwelled on the environmental aspect of sustainability – and it is a big issue. However, in the organisational environment in which you operate you have responsibilities beyond simply saving the planet. You have duties towards your paymasters – your key stakeholders – and these comprise the community in which you operate. This community may be your local residents or your tenants but they are your raison d'être and in terms of sustainability their wellbeing is one of your major concerns.

This 'wellbeing' element breaks down into two categories – economic wellbeing and social wellbeing. We will look at economic first.

Opportunity and obligation

Although this sub-heading sounds like a Jane Austen novel, it sums up the approach to local sustainability when procuring a contract. Tendering a new contract in an area always presents a new opportunity to generate local economic growth:

a) Is it a small contract suitable for a smaller, local firm or

b) If it is a larger contract, can it be divided into Lots to allow smaller firms at least a chance of securing some of it or, lastly...

c) Is it really only suitable for a larger provider?

Low Value Tenders

Small contracts – with a value below the EU Services threshold (and whether works or services) – *must* be tendered using the Open procedure. This requirement has been invoked on the back of Lord Young's recommendations with the intention of opening up the opportunity for smaller firms. Not everyone agrees this will have the desired effect, but that is the law.

Low-value tenders above the EU Services threshold may be tendered using the Restricted Procedure, if you wish.

Lots

The Regulations require you to divide larger contracts up into Lots wherever possible: if you haven't, you must explain in your OJEU why not. You can pre-determine how you will award these Lots (i.e. the allocation mechanism) and this will give you scope to reserve Lots for smaller firms by determining that they will not be awarded to a bidder as part of a bundle.

Larger providers

When a bigger provider has been lucky enough to secure a large contract, either as one lump or as a collection of smaller Lots, they should feel obliged to put something back into the area in recognition of having gained that work and the larger contract should enable them to do that – within reason.

On this point, many contractors do willingly align to this principle, but some don't. It is for this reason that realising these gains is an important part of the procurement process and not part of the "I thought they would but they haven't" approach when it doesn't happen later on. You need to *make* it happen. We can look at how.

How do we do it?

There are several ways to create local economic development within your contract and as is so often the case in procurement, there is no black-and-white answer on which one to choose; as usual, it all depends. Primarily, you have the choice of stipulation and insistence – writing what you want in the specification and the terms and conditions – or seeking the one that best meets your aims by asking for proposals within the tender evaluation model. Normally, you will end up with a combination of both, insisting on the basics and seeking the best on the 'bonus' benefits.

There are other factors you need to consider in parallel with the 'how?': what scope is there within the contract to bring economic gains? A small paving contract can in reality bring little to the table; a five-year partnering contract to refurbish half a neighbourhood's housing stock provides a wealth of opportunities. Meanwhile, there is everything in between. So the answer to 'how do we do it?' depends initially on what you are doing it *with*. If the paving contract is worth £30,000, I would not expect you to have an extensive quality submission within the tender and in fact a service of such a value would probably be sought in public organisations through a simple quotation process with no quality assessment at all. You would have no opportunity to 'seek' sustainability measures and you would have to prescribe them – normally as part of the specification or as part of the conditions of quoting.

With a larger, tendered contract like the partnering one, sustainability would be a huge part of the client's requirement and bigger stuff can be specified (in the contract preliminaries and the terms and conditions) and huge ideas can be floated as part of the quality evaluation.

A word of caution: you cannot tell a firm that they have to use local labour. To stipulate such a thing would be contra to the ideology of free movement within the EU member states and therefore against the law. You cannot tell a company not to bring their labour force with them or not to employ people from, for example, Italy. That would also be breaking the law. However, all is not lost: the local employment issue came to the European Court in the Beentjes

Case in 1987[9] and a workable precedent was set. EU law recognises that local benefits should come from a contract, provided it is not discriminatory or to the detriment of other member states' rights.

In the spirit of this, you could seek bidders' views on how they would stimulate local economic growth and the use of local labour, or you could stipulate that they must seek to fill vacancies through the use of local labour first, before seeking suitable employees elsewhere. However, you cannot expect them to employ a person just because they are local, or to employ no-one other than local people.

Similar considerations will apply to other areas of local economic growth such as the local sourcing of materials.

What can we look for?

Enough of the 'how': now the 'what'. Local economies thrive on two things: trade and employment, and these are what you need to encourage. Trade can be established through the use of locally-sourced materials and the use of local firms for sub-contract work. Employment can be created through vacancies being filled from the local labour force.

Remember, you cannot insist on any of this (see above –Beentjes) but you can encourage and expect it if it is feasible (and I repeat: non-detrimental to other member states' rights). Just bear in mind that it is good using local suppliers, but your contractor may be able to source materials considerably cheaper elsewhere. Are you willing to pay more for locally-sourced materials? If so, how much more? We come back to balancing costs against benefits and the corporate 'stand'.

The use of local firms and the development of SMEs[10] is a big issue for local authorities and a very good route to stimulating local economies – SMEs tend to employ and buy locally as well – and securing them work with the bigger players is an area where local authorities (and similar organisations) can be very pro-active.

9. See http://oxcheps.new.ox.ac.uk/new/casebook/cases/Cases%20Chapter%204/
Gebroeders%20BV%20v%20The%20Netherlands.doc
10. SME – Small and Medium-sized Enterprise. Normally defined as businesses with up to 250 employees (fte).

Such organisations can provide advice on how to secure work with the 'big boys' by organising training events and seminars and can also lay on 'meet the buyer' events where your smaller local firms can meet your larger providers to find out how to develop a trading relationship. Such events do work, and provide very good publicity for the organisation that stages them.

An organisation's major suppliers will generally be happy to participate in such events (for various reasons) but securing assurances of this support could also be part of your tender process and you may well find firms bidding who are better at setting up these interfaces than you are.

Going that step further

If you are bold and have the right partner on board, the development of the local economy and employment can take on a whole new meaning. I am not advocating this as the way forward, but a way of generating social development is through the creation of a social enterprise[11]. In the context of construction, the organisation you work for would work with your contractor or contractors to establish a social enterprise designed originally to train and upskill local people and equip them to work for your main provider. Once this is established, the concept can be developed: it can start to *supply* the operatives on a quasi-agency business and continue to develop to the stage where it becomes a *bona fide* sub-contractor to your main providers.

Whilst you the client and main contractor/s have established and driven the project forward, management is gradually handed over to the enterprise's own managers and an independent entity is established, able to tender in its own right for work with your main providers and on the open market.

The idea is that the enterprise is set up for the future, as a legacy that will grow and continue to provide opportunities for your local residents. Of course, such a proposal costs money, demands

---

11. Social Enterprise – an organisation that uses capitalistic principles for philanthropic goals, operating on a profit- or non-profit making basis.

considerable resources and requires certain skills; to make it a success you need the right commercial partner or partners on board, working with you over a long period of time but there are contractors out there who have done this sort of thing.

As an idea, then, creating a social enterprise is good (the ultimate community benefit), but in reality only fit for certain contractual relationships, but that doesn't mean you should not consider it...

The Regulations go a step further. Even if you have not set up a Social Enterprise, you can, at time of tender, reserve contracts for Social Enterprises, Mutuals[12] and Sheltered Workshops. A couple of points on this:

- You can 'reserve' a contract simply by stating this condition in the OJEU Notice and referring to the appropriate Article in the Directive (you may need some help with this)
- Sheltered Workshops operate commercially but provide work for people who are disabled or disadvantaged (although 'disadvantaged' is not defined in the Regulations).
- Such workshops qualify as Sheltered provided at least 30% of their workforce is disabled or disadvantaged and then they are eligible to bid for reserved contracts.
- Social Enterprises and Mutuals cannot re-bid for the same contract for a period of three years of having already provided that service (e.g. cannot tender at re-let)
- A contract with Social Enterprises and Mutuals cannot itself last

Social Sustainability

The three elements of sustainability can often be seen to overlap and social sustainability in particular is often achieved through a combination of the other two plus a bit more, rather than simply as a stand-alone issue. The principle of social sustainability is essentially one of enduring communities (that is, communities that last rather than tolerating communities) and two things that contribute to this

12. A Mutual is an organisation that operates on a not-for-profit basis and generally provides for the common good. All profits must be channelled back into the organisation for the furtherance of its objectives.

are a good local economy and a pleasing environment. To complete the picture, we need to add local services and facilities. A good local economy will help with this but in the world of procurement, especially when the contract under consideration is comparatively large and the provider will be a noticeable presence in the area, the bidders can be expected to commit to a degree of additional and direct investment in the community in some form or another.

This investment may be through staging events, presenting in schools, sponsoring competitions or supporting a project for a community hall or similar asset. It could progress the idea of supporting local labour and training into a training facility as a legacy, to last and serve after the contract has ended – this sounds too much to expect but it has been done and within my personal experience. That is achieving gold.

Social sustainability benefits are often the most difficult to secure and although many things are possible, you must remember to keep your expectations in proportion. – smaller contracts provide smaller opportunities and in some cases possibly none.

You are referred to the discussions on the Social Value Act in Chapter 15.

Sustainability in Context

The last paragraph above sounds trite but it is nonetheless true: You need to keep your expectations in proportion to several things. For example:

- The value of the contract – a contract worth £250,000 will provide only a very limited opportunity to contribute to the community or the economy
- The physical size of the contract – if the contract comprises fairly extensive work 'out in the community' there is a better link and therefore a stronger case for community benefits to ensue
- Type of contract: if the contract has a high labour content, more employment opportunities can be expected than from a supply contract, which would probably provide none.

- If the contract is for a long period (some partnering or PFI[13] contracts will last for many years) greater investment in the community it serves can be expected

Notwithstanding the bullets above, you should never consider a contract to be too small or in any way unsuitable to carry environmental sustainability requirements. Even a simple consultancy contract worth, say, £3,000 can have clauses regarding minimising travel, reducing paper waste, etc. These simple measures will almost certainly (i.e. should) be fundamental to your organisation's sustainability strategy and should also be fundamental to any other business's sustainability agenda, be it wrapped up in a complex policy or part of a simple statement.

Always remember the overlaps and the balances – encouraging the use of local suppliers is good for the local economy and also reduces the carbon footprint of delivery. On the other hand, you may need to seek cheaper or more sustainable materials from further afield. Your 'policy' will need to set parameters for such considerations.

Having considered what you can expect, demand or aspire to, the next question is how to achieve it. We have already considered this question but, as a reminder, you can demand your requirements in the specification and the terms and conditions, you can set a minimum standard in a PQQ and you can assess proposals as part of a tender evaluation. Often you can have a combination of all three, but remember to make sure you do not repeat questions from the PQQ in the tender.

When setting out your sustainability requirements in a tender, take great care with your evaluation model. Some benefits are not tangible, bidders may come up with ideas you never even thought of and you may have to compare a proposal offering five apples

---

13. PFI = Private Finance Initiative. Can be called PPP (Public Private Partnership). Both are contracts that use private money to secure public assets and services (e.g. hospitals, schools) and rely on a long-term commitment to recoup the investment plus profits. PFI contracts can last 30 to 35 years. More detail is not suitable for our purposes here.

with one offering four pears. You are referred to Chapter 7, but a quick tip here is to give an *idea* of what you would like to see (or what you would like to achieve or encourage) and qualify your evaluation criteria with the fact that the final decision may or will be influenced by a degree of subjectivity on the part of the evaluators.

When you have done that, test it – have a critical friend challenge your approach.

Once the contract is let, the sustainability issues become a matter for contract management so make sure the contract manager is aware of them and has the tools to monitor and enforce them. As the tender manager, ensuring this continuation is your responsibility.

Do not re-invent the wheel – I have already spoken of organisations that can offer support and in addition to these there are numerous standards that you can quote that are recognised requirements and will require little additional explanation from yourself.

Never forget – none of your requirements will come free, although they may well be sugar-coated. The cost of a company's standard environmental or sustainability policy – which you may not actually specify – will already be reflected within their tender sums; measures will not come out of profits. On this basis, the more you demand, the more it may cost. If the best benefits come with the cheapest tender sum – look into it carefully because you will have to pay somewhere down the line.

One final point – sometimes what you achieve may be small and some may question the value of these 'achievements'. Addressing sustainability within the procurement process is neither a panacea nor the route to the Holy Grail but every gain is a step – however small – that contributes to the whole. This may need to be explained to the agnostics amongst us who may not always have the 'energy' required to include sustainability, especially in situations where it may seem more effort than it is worth. Rarely can *nothing* be done and – as the saying goes – every little helps.

Sustainability is the hot topic – treat it well.

Summary

- Sustainability essentially comprises the elements environmental, economic and social.
- Do not expect more than the contract can reasonably give.
- Sustainability benefits can be gained through stipulation within the contract, requirements in the PQQ or proposals within the tender.
- Whole Life Costing can help with balancing up-front costs against eventual benefits.
- Sustainability requirements always incur cost.
- Remember to carry out cost/benefit analyses of each initiative or idea – it *has* to be worthwhile.

14 // Stakeholders and Consultation

The starting point of this book is procurement in the public, or quasi-public, sector so by definition you are probably providing building and repairs services to peoples' homes using what is essentially their (or at least the public's) money. These are two key reasons why consultation with your customers is important in the majority of the procurements you will undertake.

Another good reason is that customers probably know as well as, if not better than, you what is actually required both in terms of works and levels of service. Additionally, there is often an element of choice involved – perhaps colours may be involved – and this choice ought to be theirs. On these bases, consultation will often be a fundamental part of any procurement you undertake and care needs to be exercised to ensure it is done effectively and fairly, but without leading you into difficulties later on.

Most organisations will have established ways of consulting with residents and you may or may not have need or wish to change them. In the instance of leaseholders, consultation is compulsory and laid down in law and a prescribed, overarching methodology has to be followed.

Avoid treating consultation as an interruption or hindrance to your procurement: use it to effectively understand customers' needs, steer expectations and gain support for the outcome. However, to achieve this, you will need to understand your stakeholders.

Stakeholders

Obviously enough, a stakeholder is anyone who has a stake – or an interest – in what you are doing. Some stakeholders are easily identified: if you are externally redecorating a block of flats, all the residents are stakeholders.

Normally, though, it is more complicated than that, especially if you are tendering a large term contract for day to day repairs. In such an instance, a list of stakeholder groups would include:

- Tenants
- Leaseholders
- Housing Officers
- Leasehold management staff
- Potential providers
- Current service providers
- Council Members
- Budget-holders / finance department
- In-house Technical Staff
- Outside consultants
- In-house legal team
- In-house H&S team
- Etc.

There are probably more, depending on your organisation.

Stakeholder Analysis

It is important to draw up a list of stakeholders and undertake a Stakeholder Analysis. There are various recognised 'proper' Stakeholder Analysis models that you can study but these are often more complicated than you will need (and so probably best reserved for academic interest only at the moment). What you essentially need to do is analyse out for each stakeholder group what they need or are owed from you as part of the procurement process and what you can get from them.

The analysis should identify each stakeholder group, their ability to impact on the process and the degree to which the process will impact upon them. This latter is tantamount to an Impact Analysis[1] in its own right which may in turn go on to reveal a new range of stakeholders (if someone is affected they are a stakeholder, remember).

Once you have listed your stakeholders and their relevance to

---

1. Impact Analyses examine the effect or consequences of a project, process or event. Some organisations require an Impact Analysis as part of any project plan and will normally have their own template or format. An Impact Analysis could be looked upon as a cross between a Stakeholder Analysis and a Risk Analysis.

the project you can identify how they should be involved in the procurement process. The differentiation between tenants and leaseholders in the list above (as opposed to simply naming them all as 'residents') is because tenants have very different interests in a procurement process to leaseholders (especially if it is for a major works project – see below) and they will each need very different approaches.

On the above basis, some stakeholders for example may need to be involved heavily in the specification process (e.g. tenants, leaseholders and in-house technical staff); some may need to attend every project meeting whilst others may only need to attend on occasion. You will have to make these decisions and avoid overburdening anyone unnecessarily (you will lose their support) whilst not missing anyone out and causing offence: they may come back to haunt you...

A proper and successful Stakeholder Analysis will steer your consultation and guide you on who to invite to what meetings and how to approach them. It will help you keep everyone on board and on your side.

Stakeholder Consultation and Project Team Meetings

Having identified your stakeholders, you will need to plan your consultation in some detail, and in line with your procurement timetable. Chapter 4 on Planning the Procurement makes this point very clear. There will be key points in the procurement process where consultation will need to take place with certain stakeholder groups and there will be project meetings where the agenda will require a higher level of input from specific members of your project team (e.g. your finance, legal or technical specialists).

Remember – setting up meetings, whether they are public meetings for a large group of stakeholders or 'simple' project team meetings for an eclectic group of officers and external consultants, all requires considerable forward planning.

Meetings

Whatever the meeting, set it up plenty of time in advance – hopefully in line with a prepared schedule. Always make sure you secure a venue of suitable size and nature (do you need an IT facility or refreshments?), in an appropriate location, and have a properly prepared Agenda.

Advance notices of meetings should be sent in good time and be clear about time, date and place. Late changes should be avoided. Generate an ethos of starting on time and finishing on time and ensure that any actions arising from a meeting are recorded and that the date of the next meeting (if there is one and it is known) is made clear and agreed before you close.

Procurement project team meetings should be managed like any other properly-run project team meeting and I shall not dwell on that matter here. Consultation meetings with large groups are somewhat different and their nature will depend upon the audience. For your guidance, I shall look at the three main types of larger stakeholder group meetings – leaseholder, resident and industry consultation – and offer some advice.

Leaseholder Consultation

The difference between the processes of 'resident' and 'leaseholder' consultation is considerable so I shall look at each separately and in some detail. For each, though, there will be support from Residents' Associations, Housing Officers and your Leasehold Management Team as appropriate and you should get these groups on board as early as possible.

If you manage properties owned by leaseholders you will already know that (quite rightly) great care has to be taken when dealing with anything that involves charging them. Unfortunately, major works and day-to-day repairs all involve charging leaseholders (often quite large sums of money) and there are clear, legal processes attached to leaseholder consultation that have to be followed when procuring and managing these contracts.

The key piece of legislation is the Commonhold and Leasehold Reform Act of 2002 ('The Act'). Known colloquially as 'CLARA', the Act arose from Section 20 of the Landlord and Tenant Act 1985, and lays down specific requirements for consultation with leaseholders when you are tendering for major works and repairs contracts.

Leaseholders are obliged to pay, through the process of a Section 20 Notice, for major works carried out to the fabric of the building they occupy and the Act was designed to ensure that they are able to verify that the landlord (the freeholder – your organisation) has achieved value for money in securing a provider for the work.

The procurement of communal services are not excluded from Section 20 consultation and leaseholders would also be consulted on the tender process for the grounds maintenance and cleaning providers, for example. But for these services, lump-sum (capital) charges are not raised because service charges are levied for communal services – heating, lighting, cleaning and grounds maintenance, etc. – on a periodic (normally quarterly) basis.

Similarly, lump-sum charges for day-to-day repairs contracts[2] are not raised, because the proportion they owe of the cost of communal repairs is tallied annually and only levied if it exceeds a trigger value of £100 for their property.

Of course, the full scope of the Act's requirements is complex, with the usual legal ifs and buts, so I will just give you an overview, accompanied by firm advice to seek assistance from someone in your organisation who knows about this area of work. This person may be a legal adviser or you may have a specialist leaseholder team or leaseholder person to hand: whoever it is – use them!

Capital charges will apply when the leaseholder is going to incur costs of £250 or more due to major works. Charges arising from day to day maintenance (from your term contracts, for example), have a trigger cost of £100 per year. Below these thresholds,

---

2. Provided such contracts meet certain requirements, they are classed in law as 'Long Term Agreements' and bring certain caveats into play – often in your favour - and you should seek more detailed advice on these.

charges are not levied. Charges are normally levied in proportion to the size of the property, and this is calculated either through a system of units (points) based on the number of living spaces in the property (living and bedrooms, etc) or sometimes in proportion to the property's rateable value. In addition to their share of the cost of the works, leaseholders also pay a fee to cover their share of the procurement and contract management costs. This is normally around 10% of the works cost.

The only restriction to this general rule is when pre-advised charges apply. When a Right to Buy property is first sold, you have to advise the buyer of any planned works or services likely to give rise to Section 20 charges for the next five years. Within that 5-year period you cannot charge them any more than the amounts you have advised them, regardless of the work you actually do or how much it finally costs.

Leaseholders will always – of course – complain about these charges and this is understandable because often they are quite high, but the legal position is that you have to charge it and they have to pay it and this is your best defence when the grumbles and complaints start rolling in. The key obligation on you is to make sure that you are able to demonstrate that you have achieved best value for money. Be warned – it can be a rocky road.

Let us look at the process.

Consultation takes place at three key stages:
1. Conception of the scheme – explaining what work is proposed and why.
2. Tender outcome – giving the result of the tender evaluation and proposing a firm to whom the contract will be awarded.
3. After contract award – confirming the contract has been awarded to the winning firm identified at Stage 2 (under specific circumstances).

At each stage, you will issue a specific form of Notice:

1. Stage 1 - Section 20 Notice of Intention (NOI)
2. Stage 2 - Notice of Proposal (NOP)
3. Stage 3 - Notice of Award (NOA).

The only way you can avoid either of Stages 1 or 2 is through a specific dispensation from what used to be called the Leasehold Valuation Tribunal (LVT) and is now known as the 'First-tier Tribunal (Property Chamber)' on the grounds of an extreme emergency (urgent works relating to health and safety, for example).

The Stage 3 process is not always necessary: for example you do not need to issue a Notice of Award when you have awarded the works to a nominated contractor or if you have selected the lowest bid.

The Notices for each Stage all have to provide quite specific information, depending on the nature of the contract and the procurement route being undertaken. I will not list the various permutations here but your organisation ought to have standard templates for these Notices. At stages 1 & 2, leaseholders have 30 days after issue of the notice to come back with comments and you have to consider their input and respond within 21 days. Always, relevant documents must be available for leaseholders to inspect if they wish.

Leaseholders have the right to nominate bidders and provided these pass your corporate selection criteria, they must be invited to bid. This means that, if you are tendering off your Approved List, for example, your procurement process and timetable will have to include a provision for this additional pre-selection (approval) stage on top of the basic consultation process. If you are undertaking a full EU procurement, 'nominated' firms can simply partake in the tender process just like anyone else, so no additional allowances need be made.

The process can be time-consuming, especially if you have a lot of leaseholders, but in addition, leaseholders are getting much more 'smart': they understand the process better and have it within their power to make life more difficult than any of us would like. You will find they generally wish to curtail the amount of work being done to keep costs low, whilst your tenants will want as much work done as possible because they will not get a huge bill at the end of it. Public consultation meetings can get very interesting!

Resident Consultation

Of equal importance, but less driven by legal requirements, is 'normal' public consultation – with Residents (comprising leaseholders and tenants) – normally arranged through a Tenants and Residents Association (TRA) or a similar representative body. There are no laws governing such consultation, but your organisation will have views or ways it 'has always done it'. I shall not advocate any specific approach to this, but give some general guidance on how this consultation can be undertaken and what should come out of it.

Remember, whatever your 'resident' consultation process and however good it is, it will not give you any leeway to sidestep any of the requirements of the leaseholder consultation legislation.

Meetings for major works projects

On a major works project you would have at least one public meeting for all affected residents to explain what work is proposed and seek general input. You will find, of course, that tenants generally want everything done, and this will normally be very much at odds with the leaseholders' views and may even be at odds with your budget.

The meeting will probably be held in a local community hall and should be attended, on your organisation's part, by any relevant person. As a minimum this will normally comprise yourself (it's your project), the housing officer/s and the architect/surveyor who is/are preparing the brief and specification. The meeting will be in the evening so that people who work can attend and will probably last up to 2 hours. Have a note-taker, if you can. Tea and biscuits are always good.

This meeting will seek views on the proposals and take suggestions for other necessary works, at which point you will need to manage expectations. Display boards with pictures will always help – especially if colour schemes require a consensus - or you can have some fancy IT set up to illustrate the proposals if you wish. You can use the opportunity to seek views on colour schemes and other choices on the spot or you can leave a display of the choices for people to consider and vote on, maybe using voting slips for them to return some time later. Often, the TRA will be happy to assist in getting resident's views back to you.

The meeting is also a good opportunity to seek nominations for one or two resident representatives who will be the key link between the occupants and the project when it is on site. You should seek at least two, either one of whom should be able to attend contract progress meetings in the daytime.

This should be sufficient. If the project is complex, you may require another consultation meeting but generally the elected representatives should be able to take any reasonable decisions on the group's behalf, feeding back and taking views at the Residents' Association meetings.

If all goes according to plan, the only other public meeting will be immediately pre-contract, when the appointed contractor will be introduced and explain their proposals regarding programme, site set-up, access to properties, resident liaison (introducing the TLO) and complaints procedures in the event of any problems.

Be wary of giving too much consultation. I know this sounds bad, but if you allow too much input – which will be seen by some people as giving them 'power' – you will make your own life more difficult for the rest of the project. You will be subjected to a constant stream of requests, changes and suggestions, all of which will have to be addressed (even if it is a simple rejection). Make it clear that consultation is limited to the process and times you have specified; your notice inviting people to your first meeting can explain that this is the opportunity for them to have their input into what work is done and how.

Lastly – beware equalities issues. Remember to provide access for disabled people and additional support for anyone with sight or hearing difficulties. The Housing Officer should be able to help with this. If language is an issue, some translation service may be required, and this may be needed on the original notice and any displays as well. Know your target audience.

Feedback

Once a project is complete, it is good to have a Customer Feedback process. This is normally in the form of a questionnaire, distributed to all properties affected by the works, with a return-paid envelope attached (if the office is very local, the reply slips could be dropped in). There should be an RLO (Resident Liaison Officer) provided by you or the contractor and a complaints book on site and these will provide additional feedback on the conduct of the project.

The survey should cover the conduct and reliability of the contractors as well as details of any outstanding issues of workmanship or incomplete jobs. This feedback will then assist with contractor performance monitoring and correcting faults at Practical Completion stage. A second, follow-up survey towards the end of the Defects Liability Period (DLP) will assist with final handover.

Non-major works consultation

For non-major works procurements – for example, grounds maintenance or term repairs contracts – it is normal to seek the support of (say) one or two residents who will be able to sit as full members of the Procurement Project Team and represent the tenants' and leaseholders' views on matters relating to the specification, levels of service delivery, and so on. If the contract is to cover a multitude of blocks (often such contracts are Borough-wide), a process needs to be put in place to elect these representatives, or it may be possible to secure volunteers through a Residents' Council or similar coordinating body.

It will be obvious that public meetings for residents across a large catchment areas (for example across a whole borough) for a term contract is not feasible, so it should be made known that, if a residents' representative body (e.g. a TRA) wants information on the proposals at a Residents' Association meeting, then an officer will be pleased to attend and also residents' newsletters can help with the flow of information. Primarily, though, input and feedback is through the elected representatives.

Industry consultation

A very useful, albeit obvious, tool for anyone carrying out a procurement is asking people who actually *know* about doing the work you are tendering. Whilst procuring officers will gladly seek the input of technical experts within their organisation, for some reason they will often fight shy of seeking the views and advice of the industry itself.

This guardedness is ill-advised and should be overcome – a lot goes on in 'the outside world' that we, in our relatively insular organisations, can easily miss and we need to keep up and learn. We can do this through industry consultation, a process the new Regulations positively encourage. This consultation can take three distinct forms, any or all of which can and should be undertaken.

Asking the incumbent

If the contract is a continuation of an existing service (e.g. a term repairs contract), the contractors who know it best are those who are performing the current service. Do not be afraid to secure their input on the new procurement@ what is good, what works, what adds costs, what deters bidders, and so on. Their involvement can be as detailed as you wish (short of writing the documentation, of course) provided they gain no advantage at time of tender.

Of course, they will have advantages: (a) they are already doing the work so they know it backwards, (b) their set-up costs will be minimal as they are already there and (c) they have an insight into the new tender as they have assisted with some parts of it. However, these circumstantial gains do not mean the tender process will grant them any favours: no questions should advantage them, no scoring should take on board anything other than what they submit – evaluators must ignore anything additional that they know about them, be it good or bad – and so on. Make sure the evaluation is unchallengeable. You will then be safe and the incumbent may win – or they may not.

General consultation in the early stages

This is open-house, where you advertise through the trade press (or a PIN Notice) or by direct contact (e-mails to a distribution list constructed from directories or professional bodies) and invite firms to an event where you will explain your intentions and seek views and comments on your proposals. You can hold a large, open meeting with all invitees present or you can meet a range of providers individually. If you choose the latter, make sure you have another person with you to take brief notes of the suggestions that come out of the meeting because they can also verify that no collusion took place.

This type of consultation should be done at the very earliest stage of the tender and will provide an opportunity for you to learn of latest developments in the industry (methods, materials, machinery, etc.) and get input from potential bidders on your proposed tender process, packaging, contract lengths, and so on.

Generally, within obvious limits, the more open you are in these consultation meetings, the more open and constructive will be the industry's response: they will appreciate your seeking this advice and will know that, in turn, they should get a tender process that better-matches their needs. Achieving this 'rapport' will benefit you and them both.

Specific consultation at tender stage

Sometimes referred to as Meet the Buyer, the second type of industry consultation is held when you start to tender and is designed to advise bidders on the opportunity and explain how the tender process is going to work.

You should hold this event a little while after you have entered the first stage of the tender process. This will usually either be early in the PQQ stage (Restricted procedure) or soon after the ITT has gone out (Open Procedure), but do not hold it too early:

you want those attending to have had a chance to look at the documentation and, if using the Restricted Procedure, you want to maximise the number of firms who have seen the opportunity and sought a PQQ.

You can include a notice of the event in your tender advertisement or in the documentation you issue or both, and make the nature of the event clear. The event should:

- Explain what the contract is about: its nature, purpose, value, length and so on.
- Cover key aspects of the specification and the terms and conditions, warning bidders of some of the contract's more demanding requirements.
- Go through the tender process – questionnaire, tender return and so on.
- Provide advice on 'corporate' standards – aim at reducing the number of firms trying to bid who cannot meet your
- organisation's minimum requirements.
- Explain your e-tendering system, if you have one.

You will see that, if executed properly, such an event can serve to reduce the number of clarifications sought during the tendering process and help prevent firms bidding who are really not up to the job, saving your time and theirs. The event can be as simple or as elaborate as you choose: you can incorporate presentations and seek questions from 'the floor', although of course questions will be guarded as attendees will wish to preserve commercial confidentiality and hopefully maintain a bidding 'edge'.

Ensure accurate notes are taken (having more than one note-taker will help) because there will certainly be things coming out of this meeting that you will need to act on later. All questions and your responses must be distributed to all bidders – those who were there and those who were not. These responses need to be issued as public clarifications, serving to further reduce the number of requests for clarification you receive and to make sure that no-one can claim they were disadvantaged by not being able to attend the meeting.

Which way is best – before or during?

You cannot compare – they serve different and distinct purposes and my advice would be to do both. Remember – if you want to know about providing a service, ask those who provide it.

Summary

- Resident Consultation is a vital means of understanding the needs of the customer.
- Leaseholder consultation has specific legal requirements.
- Stakeholder Analysis should be used to formulate the consultation and communications plans for your procurement
- Consultation meetings need to be well-planned and properly organised.
- Industry consultation can be extremely beneficial.
- The Regulations advocate it.
- PIN Notices are a good way to get industry involved early on in the procurement.

15 // Services

What are they?

In the opening sections of this book I made much of 'Services' with a capital 'S'. This is because the EU makes much of it, too. Getting 'Services' wrong is one very quick way of falling foul of the Regulations regarding Ineffectiveness, and if you have read Chapter 11, you won't want to do that. If you have read everything thus far, you will know that Services have a different EU Threshold and you can – in certain circumstances – invoke the Light Touch Regime. You also have the influence of the so-called Social Value Act[1]. Believe me, those who deal in 'works' don't know what they're missing.

The Public Services (Social Value) Act 2012, in brief, requires an Authority, when tendering *any* Service, to consider what social value it can add as a benefit to the community. For some reason, this obligation does not apply to works or goods procurements. To my mind, the requirements of this Act are a little odd because, for the most part, services tend to be for social value purposes anyway, so the requirement is already in the frame; secondly, the Act only requires *consideration* of the possibilities – there is no mandate to deliver. As good procurement practitioners, we would consider social benefits as a matter of course so I see no problem or additional burden, but the Act is there and needs to be acknowledged.

So when is a Service not a Service? This question causes people a lot of bother. If you let a contract to install boilers in a block of flats, that will be a works contract and you can include a two-year maintenance agreement as well, which would be a good idea. If you then let a 5-year contract for the follow-up day-to-day maintenance of the boilers, that is a *Service* contract. Not because it is five years, but because the nature of the work is providing a service rather than creating or improving an asset.

For this reason, term repairs contracts – regardless of their value – tend to be defined as Services under the EU definition whilst a

1. The Public Services (Social Value) Act 2012

newbuild project is clearly a works contract. On this same basis, you could distinguish them by saying works contracts use capital funding and Services are supported by revenue, which is another reasonable logic. Often, though, the distinction between the two is not so clear and people have 'opinions': not all Services are provided through ongoing term contracts and sometimes Service provisions can be funded through capital grants.

Sometimes a contract will be a combination of both works and Services; the official guidance here advises that you should classify the procurement on the basis of the predominant element. There are some rules of thumb[2]:

- If it is Services and Supplies, the classification is determined by the financially predominant element
- If it involves works with either Services or Supplies, the classification should be determined by the predominant purpose
- If the contract supplies equipment and an operator, it is Services
- Supplying IT software is a Services contract if it involves tailoring the product to the purchaser's specification.

This is all fine – the assumption is that, for most of the contracts we are dealing with here, classification is primarily judged on the basis of either predominant cost or predominant purpose – the Regulations stipulate this – but bear in mind that, if you get it wrong and the resulting classification results in you not publishing an OJEU Notice when you should, you will be in breach of the Ineffectiveness Regulations. What all this means is be sure and, if you are not sure, be safe – publish a Contract Notice in OJEU anyway.

Moral: if you are not sure, be safe: always seek the advice of legal and remember that the erstwhile[3] OGC can be used as an additional source of advice, via the Cabinet Office, if you wish.

---

2. From OGC "EU Procurement Guidance" 2008
3. See notes elsewhere on the OGC's move into the Cabinet Office

Once you have decided you are letting a Services contract, you then have to decide what the procurement route is to be. The new regulations do not distinguish between the types of service but where the Service falls under certain CPV codes (as listed in Annex 14) and within a certain price bracket there is the possible option of tendering using the Light Touch Regime – see Chapter 3 on Procedures.

Tendering for Services

The method used for tendering for Services depends on the nature of the service – this seems an obvious thing to say but it is true. In the examples I have given above, some Services contracts seem to be works in everything but name. As examples I will cite you a grounds maintenance contract or a day-to-day building repairs contract. These types of Service will be based on works-type tasks and normally be priced against a schedule of rates (or standard order descriptions).

For Services contracts of this nature, the procurement is exercised just as if it were a works contract, but using the Services threshold as opposed to the works threshold to determine whether a full EU process is required or not[4]. For Services of this nature, an EU process is nearly always required. If you think it is not – double check.

On a day to day basis, the procurement will be just like a works contract and evaluation will normally be against *MEAT* criteria; the contract will almost certainly be a term contract and partnering is a good option to consider. At this point I can refer you back to Chapter 7 on evaluation techniques and Chapter 9 on Partnering.

However, the procurement of non-works-type Services requires a different approach and some specific consideration here.

Non-works-type Services include contracts for surveys, research, social service provisions such as day care, and so on. For us, the most likely Services we will need to procure will be those for construction consultancy services (QS, etc). I shall therefore look

---

4. Unless you can adopt the Light Touch Regime

now at tendering for construction consultancy services but can advise you that a tender for any type of consultancy service will be done in a similar manner.

Tendering for Consultancy Services

I have spoken a lot about construction and repairs contracts and we have looked at the use of consultancy services – quantity surveying, building surveying, and so on. Sometimes these services are available in-house but sometimes they are not and you will have to run a tender to secure the services of an outside provider. If this is the case, remember to add this time to your main contract's tender programme.

Like a normal tender, but different

Always treat a consultancy procurement as a proper procurement in its own right, not as just an additional task or bolt-on. Naturally, the larger the value of the commission, the more work it will entail but a consultancy commission of any notable size will require planning and timetabling just like any other tender and all of the preceding chapters in this book will apply. Do not underestimate the importance of getting it right and the work involved in doing so.

The appointment of consultants is often referred to as a commission and the process of procuring them as commissioning. In apparent contradiction to the comments I made in the introduction to this book, I shall observe this convention.

There are other differences, too. The specification of what services are required is called a Brief and the tender return is known as a fee bid.

The Brief can be what you want it to be but essentially it comprises the specification for the required services. For a small commission, for example a one-off resident survey, the brief might comprise just one or two sides of A4 and request a written fee quotation by the set deadline. If the commission is large – say providing QS Services for a major term contract over the next five years – the Brief will expand considerably and include the tendering documentation, terms and conditions of appointment and all other documentation relevant to the bid. A suggested list of such content is given later.

Catering for Stages

For some consultancy services you will have distinct stages. Architectural or other design services are typical examples (see RIBA Stages, Appendix 8). If you appoint a practice to provide services up to – say – Stage 4 (Technical Design) and then choose to keep them on for the duration of the project you have a problem:

i)   Their contract is finished and
ii)   You cannot simply award a new one (procurement regulations/standing orders)
iii)   If you re-tender for the rest of the project they may not win, whereupon...
iv)   Another practice will take over the design job. These new people...
v)   Will have to learn the project all from scratch and...
vi)   Who will have liability for the design work done so far?
vii)   We haven't even mentioned IPR[5] and who owns the design work done so far.

It gets messy. You should always tender for *all* stages and include the option to terminate (or not continue) for succeeding phases. The Fee Bid (see below) should be structured to reflect these natural breaks so that, if the project does not proceed, your financial obligations are clear and defined. To match this structure, always make sure *you* own the IPR of all the design work for which you have paid.

The Fee

If the consultancy is being asked to complete a project – for example a piece of research or a feasibility study – the fee will be expressed

5. IPR – Intellectual Property Rights to the designs – the right to use them as your own (which they are).

as a price lump sum. If you are seeking bids to perform, say, the QS services for a tendering process, the bid will normally be expressed as a percentage of the contract value, although a lump-sum fee is possible[6]. If asking for bids to provide ongoing contract auditing services, the fee will always be expressed as a percentage of the main contract's value.

If the consultancy is to provide architectural or design services for a project, it is normal to ask for the fees to be broken down into stages (often the Royal Institute of British Architects stages – see Appendix 8) so that, if the project halts or terminates at any time, the financial commitment at that stage is clear. With lump-sum and percentage fee bids these stage payments can be calculated, based on anticipated project sums as originally agreed but it is not so straightforward.

However you ask for the main fee to be tendered, always remember to seek, in addition to the fee bid, a quote for hourly rates for key consultancy staff members (and their supporting administration services if you wish) so that, if any additional duties crop up (for example, an additional audit), you have already competitively established the hourly costs to be applied.

Quality evaluations

You can include a quality assessment as part of your tender evaluation process or you can set minimum quality or experience requirements as part of your specification. If the fee will be substantial (for example, securing QS Services for the life of a large term contract will cost a considerable sum), then you will exceed the EU Services threshold and so a full EU tendering process will be required. In these circumstances, you will be using a Questionnaire (PQQ or BQ) and will need to include a quality element in the tender evaluation process.

Quite often, however, it will be the case that consultancy bids will be relatively small in value, when the service can be sought

---

6. If you ask for a lump-sum bid, consultants may still base their lump sum fee bid on their standard percentage fee rate.

via a Request for Quotation (RFQ). In such instances, quality elements can only be stipulated, not evaluated as part of the submission.

If the 'consultancy' service is for architects and you are holding what is called a Design Competition, the whole process becomes more complex and really stretches the remit of this book. But we will have a brief look in a moment, just so you know.

Frameworks

You can, of course, have a Framework for consultancy services – and organisations often do. You can appoint a consultant *from* a Framework or you can set up a Framework with consultants on it.

If appointing *from* a Framework, observe the rules: is it a mini-competition or do you simply call-off the commission using the percentage fee rates they have lodged as part of their original tender submission? Are further negotiations on fees allowable? Each Framework will have its own processes (see Chapter 8) and you have to abide by them – just like a works Framework.

If you are setting up a Framework, the process is similar to any other Framework tender process but:

a. You will need to decide how consultants will be appointed from the Framework – call off or mini-competition, for example.
b. You will always need to include a quality element in the Framework's original evaluation process
c. The Framework brief will need to be broad enough to encompass all the potential commissions, but this does not necessarily mean it will have to be akin to War and Peace[7]
d. You may need different panels within the framework to cater for different disciplines
e. You may have a multi-disciplinary panel.

---

7. Leo Tolstoy, 1869

Health warning: if you need to commission a consultancy for your works project, do not let mission creep 'con' you into setting up a Framework: remain focussed. If you *need* a consultancy framework, that is another matter entirely, but proceed with trepidation.

Format of a Brief

The Brief you produce will be yours to make of it what you wish, but there are some key points to bear in mind.

A small commission may require just a single page or two of A4 paper describing the Commission, along with a copy of the Contract or Agreement that will be signed, and a request for a written quotation by a certain date. As the commission gets larger or more complicated, more detail is required. Once you get to Consultancy Commissions on the scale of the EU threshold, the list of things the Brief will require gets bigger. The Brief and tender documentation for such a Commission will include all or any of the following:

- A cover sheet with clear titling
- Distinct sections for each part, all set up in the form of numbered paragraphs and sub-paragraphs. This keeps everything concise and easily cross-referred. The sections will comprise:
- An index or list of contents
- An overview of the Brief's requirements – a commentary on the Commission
- Details of the client – the operating environment
- What the consultant is expected to provide – the specification
- What the client will do in return (e.g. equipment and access provision, IT systems, etc)
- Operating terms and conditions (i.e. the contract) – or a referral to the form being used8
- Payment terms including conditions surrounding costs and disbursements
- Instructions for tendering (or bidding)

8. As explained, your organisation may have its own form, which would be included with the brief but under separate cover. A Standard Form need not be enclosed, simply referred to.

- Corporate Questionnaire, if to be included in this part of the process
- Commission-specific Questionnaire (i.e. the Business Questionnaire)
- Commission-specific tender quality questions
- All evaluation criteria and scores
- Bid evaluation scoring model – including how the MEAT criteria are evaluated
- Questionnaire and tender submission forms as appropriate
- Undertaking of honesty and non-collusion, for signing.
- Return label if it is not an electronic process.

It seems a long list, but for a larger commission you will need all this (I refer you to tender packages in Chapter 16). Try to contain everything within the one document, including the sections for submission returns. You can instruct bidders to complete a copy of the Brief and submit this as one whole document – this will ensure the bid is submitted and signed alongside the requirements of the Brief, which can then become the main part if not all of the whole contract.

The astute will realise that if all the items listed above are included, we will have prepared a package suitable for an Open Procedure tender. I have included the Corporate Questionnaire and the Commission-specific Questionnaire, the quality tender submission and the price fee bid. If you choose to use a Restricted procedure, some of these parts may need taking out and assembling elsewhere. Your choice of procurement route will determine this, so perhaps you should have another look at Chapter 3.

I have also assumed you will be issuing the Brief via e-mail and receiving hard-copy returns. If you are not, you may need to tweak the documentation and the submission process slightly. There is a section in Chapter 16 to help you with this, but remember - whichever method you use, make sure the Instructions to Tender accurately describe the process.

Design Competitions

Design Competitions are normally held when you are seeking the services of an architect. We spoke about subjective evaluations of quality submissions in Chapter Seven, and the assessment of design submissions will comprise a lot of subjectivity. That is the first thing to make clear in your documentation. Having got that straight, let us look at the rest of the process.

The EU Regulations are sticky on Design Competitions: if the competition (i.e. the work of the architect) *leads to a contract* that is above the EU threshold, then an OJEU Notice must be published, regardless of the value of the design contract itself. If the design team contract is to be novated to the constructor once the constructor is appointed (see later) the design costs will need to be included in the build cost calculation[9] and this will have to be considered within the strategies for the architect *and* constructor procurements.

The Brief

The brief should explain in great depth your requirements for the building, whether the competition is for a new build or the alteration of an existing structure. You should provide information on how far Planning have been involved to date and advise of any issues such as Listed status, conservation areas or known requirements of English Heritage. You need to list clearly your minimum requirements as a client: building capacity, access requirements, anticipated footfall and so on. You will include location maps and identify all relevant statutes and documents (such as the Local Authority's Unitary Development Plan), that might impact on the design process.

Often, the designer's contract is novated to the constructor once they are appointed because placing the constructor in charge of the whole delivery side including design services avoids issues around liability for design errors, changes to design, etc. If it is intended

---

9. Certainly the constructors will include a cost for assuming this responsibility and its liabilities.

that this is to be done, it must be made clear in the architects' brief (and in the constructor's contract specification).

The brief must also provide all contract information and instructions for bidding. You will need to explain the tender process and define what the architect's roles and responsibilities will be as the project progresses through different stages. For example: how involved they will be in the tendering process for the constructor and what will be required of them once construction works begin on site. On this basis, you may wish to break up the fee element bid into costs for certain specific stages in the design and construction process. To assist you with this, there are accepted stages laid down by RIBA (Royal Institute of British Architects)[10] and these will help you identify the level of design detail and input you will require of them at each stage of the tender and the actual construction. There are other accepted standard stages that may also be used, for example, the Landscape Institute has a similar regime to match their own particular area of operation.

Procurement and evaluation – the process
Having got your brief ready – it will be bulky and take some time to complete – you can embark on the tendering process: the 'Competition'. The procurement route is for you to decide and the normal considerations will apply – see Chapters Three and Four.

It is recommended you adopt the restricted procedure: there will generally be a high level of interest and you are advised to pre-select on grounds of corporate suitability alongside experience of having worked successfully on similar commissions in the past.

Normally, the design evaluation process will comprise two stages: an initial selection to shortlist and then a final decision stage. Because of the work involved in the final submission, it is normal to pay the shortlisted firms an honorarium to help offset the costs and for this reason you need to limit the number of practices you shortlist. Normally you would shortlist about three – or whatever number your budget will sustain!

---

10. See Appendix 8

The tender evaluation of a Design Competition will probably have to take on board four key factors:

- how well the look of the building meets your wishes
- the degree to which it meets your building performance requirements (capacity, sustainability, etc)
- the build (or whole life) costs and lastly
- the designer's fees.

How you apportion the value or weighting of these four elements in your scoring matrix is up to you, provided it is made clear at the outset.

You will need to bring together an evaluation panel of sufficiently wide experience. Whilst a QS will be essential when evaluating costs and building performance, the client should make their own judgements on a building's looks, although the input of Planning and/or English Heritage may be required as well. A surveyor would also be useful when considering a building's performance and 'usability'. Make sure you get a suitable team on board from the outset, and use their skills to help hone the brief and the evaluation model.

Whilst there are some accepted standard procedures, the evaluation process can be what you choose to make it so long as your chosen process is clear and fair. The obvious way to conduct the first, elimination round is to look at the initial proposal sketches, discard the ones you hate and then consider the other aspects of those that remain. You could then accept all those remaining that meet your budget and then finally consider the building performance aspects. This is a fairly crude process, but it would work.

You could be more technical and score every aspect – even the sketched design proposals – and arrive at a final score for the first stage and then select the short list according to your advertised criterion: maybe the top-scoring three, or all those that score over 48%.

You may go for a combination or variation of any of the above.

Whichever way you choose, it is worth mentioning that you can change the weightings of factors at each stage of the evaluation process. But why would you want to do that? It might be that your prime consideration at the first evaluation stage is the building's looks, and you intend to evaluate the performance and cost aspects with closer scrutiny at the next stage. You then know that all those designs shortlisted will *look* right so you can go on to give one of the other factors the highest weighting at the next stage. (Remember, how the looks of the building change and develop during the tendering process will still be relevant, so appearance cannot be removed entirely from the evaluation process).

If, instead of appearance, budget is the factor that has least tolerance, building costs and fees might have to have the heaviest weighting at the first stage ("I don't care how good it looks, I can't afford it") and the other aspects can gain prominence later.

Remember – whilst you can choose your method and vary weightings at each stage as I have explained, you must *not* change your evaluation model half way through the process. You must explain it all clearly at the outset and stick to it.

Assessing the design submissions

The first design-assessment stage will generally only require architects' sketches or impressions of the envisaged building, showing how it will look and serve to meet the client's visual requirements as stated in the brief. These drawings will be supplemented by written submissions confirming how it is proposed the design will meet the client's 'hard' criteria such as build costs (your budget), capacity, build-time, sustainability levels, and so on, plus finally the anticipated design fee.

To develop the concepts into final proposals, it will be necessary to meet with each shortlisted practice to talk through their ideas and discuss how they have to develop them to better meet your needs and your budget. Like Competitive Dialogue, you must not leach information from one practice meeting to another – original ideas to meet your requirements are the designer's commercial edge in the final competition.

Like Competitive Dialogue, you could have more than one of these 'design development' meetings with each bidder, but it can make the process drag on. You are better off making the design competition as compact as possible, appointing the practice and getting on with the work. The bidders would prefer that, too, and it will reduce your honorarium costs. Everyone's a winner.

Final submissions must include proper design drawings accompanied by much more detailed build costs and performance data that can in the main be substantiated. Subjectivity will still play a key rôle in evaluating the visual submissions, but hard evaluation criteria should be used to evaluate the fees, build costs and performance figures. Ideally, the evaluation process should be supported by post-tender interviews to allow concepts to be explained and details of the proposal to be clarified.

Honoraria

It is normal, because of the amount of work involved in developing designs from first principles to a state suitable for final appraisal, that bidding practices who progress beyond the first (or second – how many are you having?) stage are given an honorarium to help offset the costs incurred. The size of these honoraria will depend on the nature of the design, and this needs to be (a) carefully considered and (b) included in the procurement costs for budgeting purposes. The availability of an honorarium for participating practices should be made known at time of advert, so as not to deter smaller practices from the competition on the grounds of cost.

Remember

Timetable a Design Competition very carefully: they take a long time to complete, especially if residents or Members are to be involved in the process, and architects will need sufficient time to get their designs and their figures right.

Summary

- 'Works'-type Services are tendered just like a works contract
- For a consultancy-type Service, all the requirements are set out in a 'Brief'
- You are more at liberty to specify the quality and experience required rather than seek a quality submission
- Lowest-price consultancy tenders are called fee-bids
- Tenders for design services (e.g. architectural; designs) are normally referred-to as Design Competitions and specific rules apply.
- Design competitions normally attract Honoraria for participants beyond the first or second stage.
- In certain circumstances, the Light Touch Regime may be invoked
- Ensure you secure Intellectual Property Rights to any design work

16 // Other Procurement Considerations

What are Other Procurement Considerations?

You will hopefully by now have realised that procurement is not a simple, episodic process: it involves a multiplicity of overlapping tasks and considerations. This means that, in the interests of avoiding too much confusion or deviation (I hope I have managed that so far), some things have not been dwelt on when they perhaps might have been for fear of diverting attention away from the main issue at hand.

I have therefore decided to put all these 'things' into one chapter, where they can be considered and then picked up by you in the appropriate context. Just because these topics have been 'pulled aside' does not mean they are in any way less important: on the contrary, they are important enough to demand specific consideration and that is what this chapter will give them.

In this chapter you will find guidance on:

- Assessing Financial Standing
- Dynamic Purchasing Systems
- Electronic Catalogues
- E-Auctions
- Concessions
- The OGC, Crown Commercial Service and Procurement Guidance
- Best Value (the new one!)
- Innovation and Commercialism
- Aggregation
- Accreditations, Registrations and Memberships
- Tender packages
- Issuing and receiving questionnaires and tender documents
- E-tendering

- Portals
- TUPE
- Bonds and Guarantees
- Novation
- Letters of Intent
- Sealed contracts
- Legal precedents
- Learning more

Assessing Financial Standing

General Points about Assessing Financial Standing
The task of assessing a company's financial capability to support a contract often gives rise to animated discussion. I hope to equip you here to actively participate in any such debates.

The purpose of assessing the financial standing of an applicant or bidder is to form an opinion on their ability, if successful at award, to sustain the contract being tendered. This opinion is essentially a calculation of the level of risk the Applicant firm represents, and the decision of the client in this respect has to be final.

You should always make it clear what financial details you require (normally one or two years' certified or audited accounts) and how you will evaluate the information supplied. Additionally, you should always reserve the right to seek clarification or further information in respect of the information supplied in an endeavor to arrive at a conclusion that is both sound and fair to all parties – including yourself.

The information can be evaluated in one of two ways: (i) stipulating hard and fast criteria against which certain aspects of a company's finances will be judged or (ii) forming an *overall* view of an Applicant's financial viability in the context of the contract being tendered. Both have their advocates and their detractors:

The first approach (above) is clear-cut and leaves no room for debate: pass or fail where any one fail means the company is not considered eligible and companies can almost predict the outcome of the assessment before they even submit the required data. The second approach, on the other hand, allows an applicant or bidder the 'opportunity' to not meet *all* the declared criteria but overall to still be considered sufficiently financially sound to progress through the tender or award process and this is the approach favoured by the new Regulations, particularly with respect to small or new businesses. It also allows for glitches that arise due to the nature of the work they do (IT software companies often have accounts that might easily preclude them from meeting strict and specific criteria) or particular events such as investment in plant or development.

Whilst financial information can be easily 'evaluated' as such, it is always recommended to have someone suitably qualified in that field to at least assist in making the final pass or fail decision.

Here is a brief description of some of the more common types or areas of financial appraisal included in the financial assessment process.

Turnover
This is the most common and basic piece of financial analysis, where you define what factor the Applicant's turnover must be of the contract's total value (generally for projects) or annual value (generally for term contracts) for the company to be considered eligible for award.

Where Lots are concerned, the turnover has to be considered with respect to the sum of the value of the Lots being awarded and this may give you a steer as to how many or which Lots may be awarded to certain bidders. You must make sure your 'rules' for making such judgments are clear in your documentation

Turnover limits are set for the protection of both the Authority *and* the Applicant firm, in the hopes of preventing a contractor becoming either too dependent on the Authority's contract or overstretched and unable to support the contract, leading to the possibility of service delivery or even business failure.

Under EU legislation the legal limit for the ratio of turnover to contract value (as defined above) is 2, but a lower factor may be asked for.

Banker's Reference

If you request a Banker's Reference you will receive from the bank a brief and impersonal letter containing one of a number of standard phrases in order that the bank may give you financial details without being seen to be disloyal to their client. A Banker's Reference Ratio can be derived from the particular standard phrase and you can set a pass/fail level for this ratio. The phrases and their associated ratios are as follows:

Phrase	Ratio
'Undoubted'	1.0
'Considered good'	0.9
'Should prove good'	0.8
'Satisfactory if part of a group'	0.7
'Respectable and trustworthy'	0.6
'Trustworthy'	0.5
'Trustworthy as part of a group'	0.4
(not used)	0.3
'Would not enter into a commitment'	0.2
'Capital fully utilised'	0.1
(no reference received)	0.0

It can be a condition (pass/fail) that a correctly worded reference from your principal banker is included as an attachment with your Pre-Qualification Questionnaire (or tender) or you can, if you wish, require a letter granting permission for you to contact their bank to request such a reference, but this is the more labour-intensive option and is not recommended, at least by me.

Ratios

There are then some key 'acid tests' that may be applied, using ratios of figures extracted from the accounts. Your documentation should make it clear that it is the responsibility of the applicant or tenderer to ensure that all necessary information is provided to enable the calculation of these ratios. The key ratios are as follows:

$$\text{Current Ratio} = \frac{\text{current assets}}{\text{current liabilities}}$$

$$\text{Liquidity Ratio} = \frac{\text{current assets - stock - work in progress}}{\text{current liabilities}}$$

The Financial Ratio = Current Ratio + Liquidity Ratio + Banker's Reference Ratio

For each of these, the actual ratio required to secure a 'pass' will be determined by the tendering organisation (hence the need for the 'suitably qualified' financial support mentioned earlier). There is no hard and fast rule and the requirements will be based on your organisation's appetite for risk, the market at which the procurement is aimed (e.g. SMEs) and the nature of the contract. Whatever the ratios are, ensure they are made clear in your documentation.

Credit Agency Checks

You may reserve the right to research the company using credit agencies, but there are some points to consider – and beware of. You should state *which* Agency you will use, what information you will glean from it and how you will use that information. You *cannot* just use a Credit Agency report to determine a company's eligibility for award: you have to base your decision on turnover limits and ratios as described above.

You can use the Agency to download the latest accounts and take on board any analysis or conclusions that Agency has come to, but you must undertake your own analysis to draw your own conclusion.

Dynamic Purchasing Systems

Whilst a Dynamic Purchasing System (DPS) *could* be used for construction contracts or services, it is not usual: they are better-suited to goods or the performance of simple services.

Essentially, like a Framework, a DPS comprises firms who have successfully passed a pre-selection process qualifying them as *able* to perform the purpose of the DPS – but there will have been no other qualifying conditions (e.g. prices). This is left to the tender stage. A DPS is Dynamic because firms can apply to join at any time during its life, with no limit on this specified within the Regulations.

A DPS must be run electronically and, when tendering from a DPS, you *must* go to *all* members using an Open Procedure and evaluate against criteria laid down when the DPS was first formed. The Standstill Period applies to all awards made from the System and a Contract Award Notice must be issued.

Electronic Catalogues

These are just what they say – catalogues of what providers can offer, accessed electronically for call-off or mini-competition purposes. Again, this procedure has little use in the field of construction but can be very useful in other areas.

Catalogues are submitted by all bidders, in a specific format as part of the initial tendering process. Call off then depends on the nature of how the ensuing Framework is established:

- Further competition for contract award may be through re-submission of the catalogues
- The Framework may allow direct call-off on the basis of the catalogues submitted at time of tender, however...
- Bidders must always be allowed to confirm that the information so gleaned is accurate and up to date.

E-Auctions

E-Auctions are not a common process, especially in local authority and similar environments, but they are used. E-Auctions form part of one of the other, standard procedures whereby bidders can submit prices electronically in a 'live' competitive process. Quality can also be assessed where it is possible to assess this 'live' as a numerical score.

I include this procedure here for the purposes of information and completeness but am not going to explore it further on the basis that it will probably be of very limited interest to readers of this book.

Concessions

A concession is a contract where the provider takes the commercial risk attached to an opportunity. A day-to-day example may be a café concession in a leisure centre – the bidder tenders a price for providing the facility, runs it and takes the profit. Bigger examples would be [a] a new road bridge, where the builders are paid through the imposition – by them – of tolls or [b] the right to operate a railway service up the east coast of Britain: the operator pays for the right to run the service and retains the profit.

A concession contract cannot normally run for more than five years, except where sense would dictate that is not long enough for the operator to repay their investment. The road bridge (above) would be an excellent example of this, where it may take 20 or 30 years to recoup the investment cost of constructing the bridge out of the net proceeds of operating and maintaining it.

Before the 2014 Concessions Directive (becoming Law in April 2016) changed the approach, concession procurements were not covered by the EU Regulations. Now, however, a threshold value has been introduced whereby any Concession with a value in excess of £4,104,394 will be subject to the full Regulations.

This threshold is based on a Euro value set every two years. The last value was set in 2016 and the next will be in 2018.

Whilst the threshold may seem high, when it is divided over – say – the full five years, it will not have to be a very big Concession for it to fall above the threshold and require a full EU procurement, made public via a notice in the OJEU.

Sometimes, you may come across Concessions where the Contracting Authority provides some sort of assistance to the incoming provider. Care needs to be taken in such instances as this 'assistance' – no matter how simple – *could*be interpreted as State Aid and all sorts of considerations need to be made. I will not go into those here – that is beyond our scope – but I will say that, where this applies, seek legal advice.

The OGC, Crown Commercial Service and Procurement Guidance

The OGC

The OGC (Office of Government Commerce) used to be a prime source of information and advice on public sector procurement and in the early drafts of this book I referred to it frequently. It was unfortunate for me and the OGC that, as part of the government's cost-cutting measures, the OGC was subsumed into the Efficiency and Reform Group within the Cabinet Office. The guidance issued by the OGC is still available online as archived material and procurement advice can still be sought from the Cabinet Office, although the nature and extent of this service, at time of going to print, is not fully known.

Habits die hard. Where I have referred to the OGC and the information it issued, I have advised of the above facts, but I have persisted in referring to it by name because the information and guidance it produced is still available under the OGC banner. The Glossary provides the necessary web links to access this very useful material.

The Guidance the OGC publishes (sorry – published) tends to cater for an audience that has procurement experience and knowledge. Now that you have reached Chapter 15, you should be OK with it.

To access the OGC publications, the following web addresses will be of assistance.

- The Cabinet Office website is at: http://www.cabinetoffice.gov. uk/
- The OGC Archive website can be found directly at: http:// webarchive.nationalarchives.gov.uk/20100503135839/http:// www.ogc.gov.uk/index.asp
- The OGC Archive website can also be found simply via the original: http://www.ogc.gov.uk/ where you will be given a link to the OGC Archive site.

Crown Commercial Service
Nowadays, Guidance from the Cabinet Office is issued by the Crown Commissioning Service (the CCS) and they can be found at:

- https://www.gov.uk/government/organisations/crown-commercial-service

The CCS now issues the Procurement Policy Notes and a newsletter giving regular updates. For enquiries, they can be e-mailed at:

- localgovernment@crowncommercial.gov.uk

Best Value
In the Introduction I made much of the principles of Best Value and lamented its loss. Well – it's actually back again, risen like a Phoenix from the ashes. Well sort of. In reality, the phrase has been adopted by the Government's Communities and Local Government department (CLG) to denote a set of procurement principles designed to reflect the new age that we are living in.

The principles are primarily aimed at what the Government calls 'Best Value Authorities' – namely those authorities that provide public services (e.g. councils, police authorities, etc.) – when working with voluntary and community groups and small businesses, but as a set of procurement principles they are sound and could be adopted to the benefit (and credit) of any tendering public body.

This new Best Value, interestingly enough, to some noticeable extent does reflect a lot of what the old Best Value stood for. A link to the actual document is given below[1] but I shall summarise here, for your immediate benefit, a little of what the new Best Value advocates:

- Continuous improvement, whilst "...having a regard to a combination of *economy, efficiency and effectiveness."*
- Greater emphasis on social, economic and environmental value in a procurement rather than simply the direct benefits of the services themselves (small 's').
- Extensive consultation as part of the procurement exercise (see also Chapter 14)
- Having due regard to the authority's (or organisation's) obligations to voluntary and community sector organisations and be responsive to their needs.
- Avoiding making decisions that have major adverse impacts on the voluntary and social sectors and SME providers. Where such actions are necessary (e.g. making severe cutbacks), steps should be taken to minimise the impact.

This, to me, is all very reasonable and could provide you with some additional leeway to reap increased social or economic benefits as part of your procurement, even if it does incur some additional cost.

Innovation and Commercialism

Best Value (above) talks of 'Challenge' and this leads us nicely onto innovation: the drive to find new and better ways to achieve the required outcome – and more, if possible.

There has been a lot of advice in this book on risks and how to avoid taking them, but context is very important: good procurement

---

1. At the time of writing, the document in its entirety can be found at: http://www.communities.gov.uk/documents/localgovernment/pdf/1976926.pdf

is smart procurement and smart procurement always looks for different and better ways to achieve better things. There is now a much greater emphasis on commercialism in procurement, even in local authorities, driven by the need to save costs and even generate income.

Risks are avoided by being 'safe' but, unfortunately, 'safe' does not sit happily in the same stable as innovative or ground-breaking and in real life you will probably (i.e. almost certainly) have to balance taking a risk against potential gains.

The degree of risk you take will depend on certain factors, such as:

- The assessed level of risk
- The level of risk you are prepared to accept (or take)
- The level of risk your organisation is willing to take
- All these balanced against the potential gains

There is no one-off answer here: it will always be very project- and organisation-specific but all the while be aware of two key considerations:

a) Procurement is in the vanguard of saving costs and creating commercial benefits
b) You are not playing with your own money – it is public money and you are a custodian of the public purse and must act accordingly.

In practical terms, if the question of balance arises, share the problem and, where possible, seek approval for any risk you deem wise to take in the interests of innovation and benefits: beware of gaining a reputation for being 'maverick' but for goodness' sake avoid getting a reputation for being 'dull' and stagnant in your approach to procurement.

Depending on the nature of the contract, ways in which contract costs could be reduced might include:

- Aggregating several smaller contracts into one (The EU Regs require you to Lot them: that can still be done, but issue in one procurement with multiple-Lot discounts)
- Scrutinising the requirements – is it all *really* necessary?
- Ensuring materials and or service levels are not over-specified
- Minimising ongoing maintenance and service costs through clever specification?
- Asking if a longer contract period would encourage better tender prices?
- Approaching the right market
- Ensuring the contract and its terms are not more onerous than necessary
- Simplifying – complications cost money
- Minimising the use of 'Urgent' delivery deadlines
- Paying the cost for *actual* delivery times when an ordered deadline is not met
- Charging administration costs for the issue of Default Notices
- Claw-backs for under performance against KPIs
- Setting up set delivery routes and times to minimise vehicle runs and maximise productivity, which may lead to...
- Fewer, albeit larger, vehicles – reduced standing costs
- Supporting a change to (for example) LPG-powered vehicles
- Investigating and ameliorating supply-chain costs
- Monitoring raw materials prices – where relevant
- Purchase control – is this order really necessary?
- Negotiating additional discounts when quantities exceed those anticipated
- Negotiating additional discounts for utilising contract extensions
- Negotiating additional discounts when benchmarking the market shows the service to be overpriced – termination and retendering is an option (depending on the contract wording).

This is a considerable list of cost-reducing tools, but it is by no means all of them. Be alert and always consider how the contract costs can be driven down without driving quality below the required level and then ensure the contract management tools are there to support their implementation. Risk and innovation can be balanced, but no-one said it is easy.

Aggregation

I have placed Aggregation immediately after Best Value and Innovation because it has relevance. I spoke in earlier chapters of increasing efficiency and obtaining better prices by combining projects as different Lots or Packages into one tender. Tender costs are reduced and economies of scale can be achieved in the tender prices.

The EU takes a view on combining opportunities too, but from a slightly different angle. The EU is very wary of organisations breaking down contracts into smaller 'lumps' to avoid the Procurement Regulations; they are also very aware of organisations deliberately *not* joining up opportunities when they ought.

This joining-up is known by the EU as *Aggregation* and it can be a sticky issue. In simple terms, if you tender a series of contracts that the EU consider to be so similar that they should be joined up, or Aggregated, they can call foul. The rules state that the values of these similar contracts should be added together – or aggregated - to see whether or not together they exceed the EU threshold. If they do, they must be let as one large EU contract using a process that is fully compliant with the 2015 Regulations.

Conversely, it is explicit within EU Law that to *dis*aggregate a contract, in other words take a large contract and deliberately break it down into smaller parts, is also illegal if done to avoid compliance with the EU Procurement Regulations. Be aware – if you disaggregate or break down a contract for perfectly good reasons and the EU Courts take you to task over it, or a contractor challenges you through the courts, you may have a very difficult job proving that your intentions were really perfectly honourable.

There is, however, a slight complication. The 2015 Regulations recognise that Aggregation can put contracts out of the reach of smaller providers and this is contrary to the EU's aims to better support smaller businesses. To counter this anomaly, rather than allow disaggregation and so a procurement outside of the rules, the Regulations 'expect' (my word) any suitable contract to be divided into Lots to allow smaller providers the opportunity to tender. We have looked at Lots elsewhere in this book, so that should not present a problem, but just be aware of it.

There is a degree of pressure attached to his. If you feel the contract cannot be divided into Lots, you have to explain why in your OJEU Notice. My view is that this requirement may have to 'settle down' and I suspect some legal precedents may have to help clarify when a contract may be considered unsuitable for turning into Lots.

The rule of thumb is: aggregate where you can, divide into Lots where appropriate and avoid disaggregation at all costs.

This may not seem to be an issue, but sometimes it is. People you work with may be loathe to run an EU tender (this is a common malaise) and persist in letting small contracts for the same thing all over the place when they really ought to be joined up and put into one, larger tender exercise. EU and financial logic scream out but do not always win and the chances of a challenge by the EU – or perhaps by an irritated larger provider - become a reality.

As an aside to this, there are arguments running (albeit on a fairly academic level at the moment) that advocate even operating an Approved List falls foul of the Aggregation rules: the process lets a series of smaller contracts for similar works through a List established outside of any EU process. The pros and cons of this argument are not for here, but can make for good water-cooler discussions – if you are of that ilk.

On a closer-to-home basis, all organisations will have a Procurement Code, Contract Standing Orders or some other internal, regulatory regime governing the tendering process. Disaggregating contracts to avoid compliance with these rules will often be deemed a disciplinary offence – so it is not only the EU who cares.

Just be warned: if you are in an organisation's procurement circle you may well come across a situation that will warrant consideration of the Aggregation issue. When this happens, you will need to advise caution and be careful that you do not become committed to something that may ultimately prove to be illegal.

Accreditations, Registrations and Memberships

What are they?
If you are not on familiar ground with this, you need to take care when requesting or assessing so-called accreditations (or registrations or memberships), because some of them aren't what they seem and are of little value in an evaluation model.

On the other hand, some registrations are legal requirements and providers *have* to have them: for example, no person can carry out *any* gas work unless they are registered with Gas Safe (it used to be CORGI), so if the contract requires this discipline, registration with Gas Safe has to be a minimum requirement. NICEIC[2] registration is a similar requirement for electrical work.

Registration by these bodies requires training and assessment and operatives are regularly monitored and inspected. These requirements can be cast-iron criteria against which an evaluation can be conducted and can make easy pass/fail questions.

Some bodies – so-called professional bodies or institutions – are joined by application, and require a minimum level of qualification in the 'trade' for a person to be accepted as a Member. Examples of these would be RICS (the Royal Institute of Chartered Surveyors) or ICE (the Institution of Civil Engineers), and members would have the appropriate letters after their name (in these cases, MRICS or MICE).

Such institutions are honourable organisations and people in the relevant profession cannot normally practice without being members, so the 'accreditation' is a good minimum requirement of individuals when tendering: the institutions demand minimum

---

2. National Inspection Council of Electrical Installation Contracting

standards of professional behaviour and will eject anyone who breaches these standards, but they do not normally monitor or 'refresh' membership, other than by demanding the annual subscription.

Finally, there are Associations, Federations and other similar bodies that cater for virtually every trade and they all have varying levels of value when assessing a firm's standing or professionalism. Some of them vet applicant firms whilst some just accept the annual subscription but add no real credibility to their members' standing or professional capability. You will have to make your own judgement about how you will value such memberships and what you will specify or evaluate in your tender.

The best route is to take the advice of technical officers within your organisation because they will know what accreditations have value and will serve your quality evaluation process.

Tender Packages

This does not mean packages that are sensitive; it means the documents you get together to send out or issue at different stages of a tendering exercise. There are three sets or packs that you need to know about, and these are sent out at:

- Pre-qualification stage (when using the Restricted Procedure)
- Tender stage (Restricted and Open Procedures)
- Actual Contract stage for signature (All procedures)

Always check whether your organisation specifies what goes out when and if they do be wary of changing anything. The documents will need to cross-refer and any changes will put this referencing in jeopardy. If your organisation does not have a 'rule', there are some suggestions below to assist you in collating what needs to go out.

Bear in mind that what you issue will depend on how you compose the documents: some of the listed items may not be separate documents at all but wrapped up together, and some may have different names to the ones I have used. These lists are in no order of importance and *all* packs should contain a list of enclosed documents and have a check list of what needs to be returned as part of any submission. The principle is to make the whole thing as logical and simple as possible for the bidder; if they misunderstand, problems will certainly ensue.

Document packs
The document packs issued at each key stage are as follows:

Pre-qualification stage (Restricted Procedure only)
- Corporate PQQ.
- Supplementary PQQ.
- Commentary on the client and the contract to be tendered.
- Guidance and instruction on completing the questionnaire/s.
- Evaluation criteria, scores and weightings used for assessing and scoring the Questionnaires.
- Undertaking of Honesty and Confidentiality requiring the applicant's signature.
- Draft specification (sometimes and if available – it can help interested parties decide if they want to bid).
- Draft contract terms and conditions (or Amendments to a standard form) so bidders know the rules they are signing up to.

Tender Stage
If Open Procedure:
- Corporate Questionnaire
- Supplementary Questionnaire (if required)
- Commentary on the client and the contract
- Guidance and instruction on completing the questionnaire/s
- Evaluation criteria, scores and weightings used for assessing and scoring the Questionnaires

Then, for Open or Restricted Procedure:
- Further commentary on the client and the contract
- Invitation to Tender (the ITT)

- Full Specification[3]
- Agreement or Contract form – terms and conditions[4] or...
- Amendments to the Standard Contract Form (if applicable)
- Contract Preliminaries including Particulars, Recitals and Articles[5]
- Form of tender (a legal statement of tender signed by the bidder)
- Tender return form[6] (your document drawn up for their actual submission)
- Instructions (guidance and requirements) for tendering
- Copy of any Bond or Guarantee document
- Evaluation criteria, quality:cost formula, quality scores and weightings, etc.
- Request for TUPE information (see later)

The Contract stage – signing up.

This is what the successful bidder and the client sign. It will comprise:

- A schedule of all the documents forming the contract
- The Agreement or Terms and Conditions or Standard Form of contract with amended clauses duly altered or ...
- Amendments to Standard Contract Form
- All preliminaries and particulars
- Copy of the ITT and the bid they submitted
- Copy of the PQQ submission (not always included but their submissions do become a contractual commitment)

---

3. The format of the specification will depend on the nature of the contract and may also include documents such as a Bill of Quantities or Schedule of Rates to be completed by the bidder as part of the tender return. See Chapter 6

4. If using a standard off-the-shelf form of contract, e.g. JCT or GC Works, there is no need to enclose the document: bidders can refer to the standard document themselves. You would just need to enclose your amendments. An actual amended Standard Form (i.e. clauses crossed out or changed as appropriate) can be used at contract signature stage.

5. These bits lay down basic facts about the contract and those connected with it but do not prescribe conditions or deal with performance. Preliminaries include details such as the parties (signatories to the contract, not wild nights out), insurances, liabilities, payment regimes, Bonds, site details and so on.

6. As advised earlier, this and the Form of Tender may be in one.

- Copies of all clarifications issued at time of PQQ and tender
- Minutes of post-tender interviews
- Minutes of pre-contract meetings
- Bond or Guarantee Forms
- And so on.

This last list has to be the least definitive as it should comprise *anything* that has changed hands relating to (a) what the bidder is required to do and (b) what they have said they will do. These are all legal commitments and will stand so long as the contract is valid.

I have not gone into too much detail about the documentation for the actual contract, here, as I do not feel this is necessary. You should be guided by your legal team on this, and they will ensure the correct documents are appropriately compiled. These contract documents will, however, have to be completed with details specific to the contract and you will either need to do this yourself or ensure that they are completed.

Hard copies – or not?

As the world strives towards a paperless environment and e-tendering takes off, it is not unusual for documents to be in electronic format, but there are limitations to this. Contracts need to be signed and usually, whatever is held electronically, a hard copy of the signed contract will be required. It is acceptable for the supporting documents to be stored on a non-rewritable CD, and this is signed and its contents listed on the hard-copy sheet. Courts will accept such a CD in the case of any dispute but do not always accept a signed CD in place of the actual signed contract.

If you have an e-tendering system, all the documents will be soft copies and you may only need to print one or two key pages for contract signature and archiving. However, fully electronic contract documentation is gaining ground and e-signatures are becoming more and more common, where *no* hard copies are required at all.

The Regulations require most public bodies to be fully electronic by 2018 so the time is fast approaching when there will be no paper copies of contracts at all but, until your organisation is in that place, it will be quite normal – and safest – for hard copies of all the contract documents to be retained and archived (normally off site) with e-versions retained for day to day use.

Documents from and relating to the unsuccessful applicants and bidders - those accumulated during the pre-qualification and tender processes – are usually retained for about a year (the safe limit for any enquiry or legal challenge or enquiry by an unsuccessful bidder) and the e-system is usually considered adequate for storing such non-contract documentation and no more needs to be done save deleting it when the time is considered right.

Issuing and receiving documents

I have dealt with issuing documents in the relevant chapters but feel it needs a more co-ordinated explanation, as it can be confusing. There are essentially three ways to issue and receive documents:

- Hard copy
- E-mail
- Electronically (i.e. via an e-tendering system).

Hard copy literally means documentation in paper format, delivered by surface mail or courier. This has probably been almost entirely eliminated by the advent of e-mail and e-tendering but it does still have its place.

I shall first work on the basis that you have e-mail but no e-tendering system and will look at each document exchange stage of the two main procurement processes from that perspective.

Restricted – Expression of Interest stage

The OJEU and / or other advertisements will advise the market how to request the PQQ: normally, this will be via e-mail to a given address. Once the advert declares the opportunity is available, the documents must be ready to go.

It is very rare that a PQQ cannot be issued as an e-mail attachment, so a template courtesy e-mail, with the PQQ attached, can be sent off immediately by reply. If it cannot be sent via e-mail you must first ask yourself why it can't and then issue the PQQs via first class surface mail, advising the Applicant via an e-mail reply that this has been done. They can then chase you if it doesn't arrive.

You then need to add the Applicant's e-mail address to a special Distribution List in your contacts. As requests for clarification come in, you can distribute the responses using Blind Copy to all Applicants on your Distribution List. Remember to use Blind Copy (Bcc in your address bar) so that confidentiality is maintained.

The return of the completed questionnaires should be possible via e-mail and this is perfectly acceptable. Remember – PQQs are not a tender, they are a submission of facts. If you wish or have to have questionnaires back in hard copy format, make this clear in the PQQ Guidance and include a return label in the package you issue (see below).

I would advise storing all e-mails with attached questionnaires in a specific folder and not to open any before the submission deadline has passed – Applicants can always ask for them back for additions or alterations and you should stay remote from any risk of being accused of involvement with any changes they may make.

To distinguish between e-mail requests for clarification and submitted PQQs, ensure your PQQ Guidance lays down clear instructions for distinct wording in the Subject field of the e-mail header. Sometimes, e-mail addresses are established specifically to receive these submissions and that is a much better solution.

Once the deadline has passed, you can either open and print all submissions (not recommended) or open and load them into a folder suitable for all assessors to access. How you do this will depend on your organisation's IT set-up. Sometimes, loading it all onto Memory Sticks is the only solution.

Once all submissions have been evaluated, you can collate all assessments into one master copy that will serve to bring all scores and comments together to assess the final outcome of that stage.

Restricted – Tender stage

If the tender documentation is compact, it can be issued as a single or multiple attachments to an e-mail. It can be zipped to do this, but Zipped folders need to be in a format that all bidders should be able to unzip, and this is not always easy. Some firms (especially smaller ones) seem to buy odd Zipping software that does not work with anyone else's, for some reason.

Always attach the ITT letter – do not make it part of the e-mail (it needs to be more formal than that) – and just use the e-mail as a note of advice that the ITT and documents are attached, with a reminder of the Tender return date and time.

If the bundle cannot be attached to an e-mail (maybe because of A1-size, scaled drawings or sheer bulk) then an e-mail can be sent to advise when and where the documents can be collected. Each bundle is then packaged up and clearly labelled as explained in Chapter 3. Always send your e-mail with a 'delivery' and 'read' confirmation. If you do not get any confirmation (and, if you wish, anyway), ring the company concerned and tell them that the documents will be ready for collection.

Tender returns must not be made via e-mail. You cannot process them in the controlled, secure and confidential manner the process requires. You should always seek hard-copy tender returns or – if you are dead set on electronic versions – ask for submissions on a disc or a memory stick, but have them delivered in an envelope! Again, you will need to issue an addressed envelope or return label for the purpose.

Open Procedure

All the ideas, suggestions, caveats and points for consideration above apply to the Open Procedure, but there is only one request stage and one return stage to deal with. Normally, requests for the tender documents should be made via e-mail and they should be issued via e-mail, if they are not too bulky. If needs be – issue them as hard-copy bundles and always demand returns in hard copy or disc/memory stick format.

Return labels

Some organisations issue addressed envelopes for tender returns. If you issue documents by e-mail, envelopes still have to be posted anyway (I don't see much sense in that) and if the submission is too bulky, the envelopes have to be clumsily stuck as labels onto a parcel. Issuing tender return labels for hard-copy returns avoids these problems and is cheaper – you simply include them as part of the e-mailed documents. The label or the front of the envelope should have printed on it, as a minimum:

- The return address as the main part, in clear, bold writing.
- In a panel, normally at the bottom:
  - A statement such as "Tender return for.....Contract" (or you can use a tender reference number)
  - The date and time of the return deadline
  - A box for signature: "Received by:"
  - A box for: "Time received:"
  - A box for "Date received:"
  - A statement such as: "Any unauthorised marking on this label or envelope will invalidate the bid"

You can add your organisation's logo if you wish, but otherwise leave the label as plain as possible.

You can use a simpler label for PQQ returns if you wish, as these do not need to be quite so formally recorded.

That's it, essentially. What I have explained above works well, but you can vary the process if you like; just remember when you need to maintain confidentiality and probity and ensure you do not change anything in such a way that it can leave you open to accusations of collusion.

The method, once you have got the hang of it, that overcomes all the problems we have just looked at, is e-tendering. I shall look at that in more detail now.

E-Tendering

What is it?

E-tendering is not strictly on my agenda and it cannot be 'covered' in a part of a chapter but I shall spend a little time on it because the government is insisting that all public bodies adopt it for all procurement exercises and so you need to know about it.

There are various e-tendering systems on the market and some are, of course, better than others. I have no intention of recommending any particular type or version but merely give you here an idea of what they could and should offer to the procurement process.

A popular misconception is that e-tendering systems do the tendering for you. This is really *not* the case, but they do *facilitate* the tendering process by removing an awful lot of paper and many administrative burdens and they can also resolve several issues around tender process security and confidentiality.

In essence, e-systems store and handle the paperwork: they replace contract files on the shelf and do the work of the postman. *You* need to carry out the procurement process, but using the e-system as the tool to store, deliver and receive the documents.

If used to its full effect, all the documents relating to a tender can be stored on the (e-tendering) system and all persons involved in the project can access them as appropriate. To monitor this access, an audit module will log all events and actions relating to all documents by all parties.

Documents are loaded up and held on the system and, when ready for issue (at, say, PQQ (OJEU publication) or ITT (tender) stage), you can set the issue time and at that precise moment applicants or bidders will be able to access them. When the bidders are ready, they can upload the completed documents on the system ready for the submission deadline yet still access them for changes and additions until the system's deadline guillotine prohibits them any further access. You will then be able to access and 'open' the documents. This process is secure.

How can it help?

The biggest benefit of an e-tendering system is the reduction of administration time:

- Providers involved in a tendering process can (must) always contact you, the lead officer, through the system.
- The system will enable you to contact all providers involved in a tendering process, either en masse or as individuals.
- This makes the whole process of responding to queries and clarifications, and advising all interested parties, a piece of cake.
- All communications are recorded, permanently, on the system.
- The system can store and use templates for the issue of Invitations to Tender, Alcatel Letters, and so on – normally as many as you choose to load.
- Late submissions cannot be uploaded by bidders.
- No paper documents to bundle, issue or receive so...
- No documents go astray in transit and no courier breaks down en route.
- No paper files on shelves.
- Access to documents is available to all officers who need it, yet...
- Nothing need be stored on your personal computer system or server (systems are normally supported by a remote server).

In addition, a capable e-tendering system will:

- E-enable Questionnaires, Forms of Tender, Evaluation Models, etc
- Link to invoicing and payment systems, so providing an audit trail for the whole tendering and commissioning process.
- Be usable for the contract operation stage as well, providing contract management support including recording payments, maintaining performance data and so on.
- Manage Approved Lists.
- Operate your own and third-party Frameworks.
- Issue OJEU Notices, etc.

This all sounds very good: is there *nothing* bad about these systems? Well of course there is.

What is the downside?

Firstly, if the system is not good, it will make tendering processes harder rather than easier – an organisation needs to be very careful about the system it chooses, making absolutely sure that it does all that is required of it.

Secondly, the problem is people. People in an organisation may be reticent to use a system which is outside of their comfort zone. A major culture-change will probably be required for those who will be involved in any part of a new e-tendering process.

Thirdly, small businesses and third sector organisations will often find such systems hard to use and may even be deterred from tendering. Bigger companies will have no trouble with e-tendering – they will probably be better at it than you are - but some small firms may actually find it impossible: they are often tradespeople who do not spend their time working on computers. There has to be a strategy in place to assist these smaller businesses and this assistance needs to be ongoing.

Fourthly, even if the system is a good system, there will need to be a process of introduction and training in the use of 'yet another software system' and *all* staff involved in tendering (and perhaps contract management) will need to attend this training.

Finally, another plus. Do not forget that if you are able to offer a full, on-line tendering facility, the EU Regulations allow you to reduce the time for key stages in the procurement process. You are referred to Appendix 7 for details of this, but remember, simply sending a document or two by e-mail does *not* constitute use of an e-tendering system!

Portals

I need to advise you on Portals and here seems as good a place as any. A search will show you various on-line facilities for reaching the market when seeking bidders and we have spoken of

Frameworks as one of these in Chapter 8. But portals are different to Frameworks in that they are more akin to a dating agency than anything else.

The principle is that when you have an opportunity you can advertise it and seek bidders through a Portal. Portals generally enable you to select bidders – often by size, type and location – and can also e-enable you to run pre-qualification and tender processes through their systems and publish OJEU Notices for above-threshold opportunities.

I use the word 'can' because there are no fixed rules about Portals and the way they operate, and some (of course) will be a better fit for your purposes than others, so you need to carry out your own investigations. As an example of a Portal, I will give a brief overview of CompeteFor[7].

CompeteFor was originally established as a route to enable SMEs to access work on the London Olympics site but it now has a much wider remit, providing opportunities for firms to tender on a broader market. Firms can sign up for consideration by simply registering themselves free of charge on the portal and potential clients can search for registered companies via the CompeteFor database against their preferred criteria as described above. CompeteFor facilitates an on-line PQQ process and through this a tender list is established. It also enables registered firms to search for opportunities.

This is not an advert for CompeteFor, but you would be well advised to visit the London Development Agency website at:

http://www.lda.gov.uk/projects/compete-for/index.aspx

and the CompeteFor website itself at:

https://www.competefor.com/business/login.jsp

where you will learn more about CompeteFor and how it operates.

7. CompeteFor is funded by all nine Regional Development Agencies in England and the devolved administrations of Northern Ireland, Scotland and Wales.

Whilst I have offered CompeteFor as an example, there are a multitude of Portals available – some National and some very regional. They can be very useful, for clients and providers alike, but you need to do your homework and make sure that what a Portal offers and the way it functions suit you and your organisation and will not lead you into contravening any of your internal regulatory regimes. And remember – Portals do *not* circumvent the EU regulations.

To investigate a Portal's suitability, they will all have a contact facility through which you can make specific enquiries, and you can always search the portal for tender opportunities and contact the advertising client for direct feedback on how the Portal works and how well it would suit your particular needs.

Portals are *a* solution, not *the* solution, and deserve consideration.

TUPE

Affectionately known as 'Tewpee', the "Transfer of Undertakings (Protection of Employment) Regulations 2006 as amended by the "Collective Redundancies and Transfer of Undertakings (Protection of Employment) (Amendment) Regulations 2014" (i.e. TUPE) are designed to protect a person's employment when the provision of a service is transferred from their firm to another – in other words, their firm does not win the contract next time around. Replacing the original 1981 regulations, the Act/s reflect/s EU directives to protect employment and is very complex and a detailed consideration is certainly beyond our scope here. Considering its impact on your procurement *is* within scope, however.

In general, TUPE applies as follows:

- If you have a service provided by a contractor and you are re-tendering this service, TUPE may apply and it probably will.
- If the existing provider loses the contract, TUPE will apply where staff, previously employed on that contract, have no

other work and so are eligible to transfer to the new provider when they take over, i.e. go with the contract.

- TUPE will *not* apply if the outgoing provider has other work for the staff that is currently employed providing your services.

It follows from this that *some* staff may by subject to TUPE transfer to the new employer.

- To be eligible for transfer, a member of staff must spend 50% or more of their working time on your contract.
- When they transfer, their employment rights transfer with them – salary, bonus and overtime rights, holiday and sickness entitlements, etc.
- Pension rights do *not* transfer.
- Although this does not concern you as a tendering body, it is a fact that transferred employment rights are not protected for ever: they can evaporate but the law is not definite about when – specific case law applies to specific circumstances.
- TUPE will not apply to new services, projects or 'existing' services when re-let in such a way that it is in law a different contract. You may need legal guidance on this last point.
- TUPE will apply if you are tendering a contract for a service currently provided by in-house staff.

Process

So far as your procurement is concerned, you have certain things that you need to do if you are tendering to continue an existing service. First of all you need to make it clear to potential bidders, in the contract description, that TUPE may apply. Note 'may' – this is important. You do not have a legal obligation to advise the tendering parties on TUPE: you have to advise them if you think TUPE may apply or not and they have to make the decision for themselves. You will think it probably applies if checking the relevant bullet points above indicates that it does. If the incumbent provider advises you specifically that they are keeping all of their staff, and says that none will transfer, then you can give the advice that you understand TUPE does *not* apply and hope they don't change their mind later when 'the other' contract they were expecting does not materialise. If there is a change, you may have to seek revised tenders. If you do advise that TUPE may apply, however, but a firm does not include for it in their bid, you cannot go back to them for a revised sum.

If the existing provider advises you that any of their staff will (or may) be eligible for TUPE transfer, you need to advise all bidders of this and explain how they should request the relevant information.

This information is supplied by the incumbent provider/s and contains details of all the staff expected to transfer over and their employment details, including pay rates, bonus schemes, contracted overtime, days off sick and so on – in other words, *all* the information relevant to the cost of employing each person. To ease this part of the process, you should make sure your contract terms and conditions always make submission of this information mandatory upon request.

You are not responsible for the accuracy of this information but you do have a duty of due diligence insofar as you need to be able to say that, so far as you are aware – or can determine – the information provided looks about right. This is uneasily vague, but if your contract is worth, say, £300,000 per year and they say they have 250 employees on an average salary of £40,000 per year who are eligible for transfer, you can surmise that this is not accurate information. Some cases may not be as clear cut as that.

Why would it not be accurate? Simply because the more expensive the staff eligible for TUPE transfer, the more the other bidders will have to increase their own bids to cover the cost of taking them on: if the incumbent exaggerates these costs, the better their own chance of re-winning the tender. If you feel there is a risk of such skulduggery, you will need to take steps to mitigate it. On one large contract where I felt such a risk existed, I paid our QS to visit the incumbent providers and check – so far as they were able – that the TUPE information supplied was a fair and accurate return. It was.

Of course, the list may change at any time as staff come and go, so it has to be a question of reasonableness and the final list is the one that pertains on the day the old contract finishes. It is normal for bidders to identify the costs due to TUPE so that, once the final transfer list is known, the tendered TUPE sum can be relatively easily amended to suit any last-minute changes.

The TUPE information has to be with you as soon after the tender goes out as possible. For the other bidders to receive it, they have to request it using a signed undertaking of confidentiality drawn up for the purpose[8] – the information will contain very personal details about individuals and data protection rules will apply. *Do not* release TUPE information to any bidder without they have returned this undertaking, duly signed. Your legal department will certainly have their own version to hand, and you must send it out as part of the tender document pack (with an explanation of the TUPE process).

It is not unusual for TUPE transfers to be problematic so always have legal on hand to guide you through the process. I will not bore you with reasons or examples, but simply advise you to allow as much time as possible between the award of a contract and start on site so as to get last minute TUPE issues resolved. I have always tried to allow 3 months for larger contracts. If you start early enough and timetable properly that is possible. You may not have that luxury and if not, rate TUPE issues higher in your risk assessment.

Bonds and Guarantees

Your risk assessment will always identify a company going out of business as a risk at the operational stage. The implications of this are considerable – the service lapses and costs are incurred when securing another provider; if the lost service is one of your legal obligations, emergency cover may be required until a new provider is appointed through due process and all of this costs money.

As a safeguard, it is normal to ask firms to provide either a Bond or a Parent Company Guarantee and you should always make this

---

8. The Legal Department should have this document.

requirement clear at time of tender. You cannot suddenly ask for it later.

Bonds

A Performance Bond is like an insurance policy, specifically drawn up to provide cover in the event that a company can no longer provide the contracted service. Bonds cost money- normally about 10% of the Bond value – and the client will pay for this, either as a directly identified cost or absorbed within the tendered sum.

A Bond is designed purely to cover the costs you incur rectifying the situation – it will not cover the cost of completing the remaining contract. You can claim the additional cost incurred through using an interim provider and the costs of re-tendering and these costs have to be ascertainable (i.e. real and provable). To meet these costs it is normal to ask for a Bond with a value of 10% of the contract sum. If there are reasons why this figure ought to be higher, it can be so, but it will cost more.

Bonds are requested depending on the value of the contract, the level of risk in it and the nature of the contract itself. You would certainly ask for a Bond for a high-value contract and one providing essential services (to pay for instant cover). You would not normally ask for a Bond for a term contract (there are various reasons for this including the fact that you will generally be holding more than one month's turnover in your hands as part of the payment process and a back-up provider will normally be in place, established by you as part of the tendering process) but this is not a rule. You always need to check what the corporate line is on seeking Bonds: organisations will normally stipulate when they *always* require a Bond and you can decide if you need to lower this bar for your own procurement.

Bonds are either 'on demand' or 'Conditional'. You will rarely come across on-demand Bonds – they are normally used in very large international contracts and in the oil and gas industries. Conditional bonds are used in the construction industry, can only be called on in the event of a default by the contractor and normally require documentary evidence of losses to support any claim against them. Sometimes legal action through the courts is required to secure the due proceeds but, so long as the Bond is from a reputable provider, it will serve to cover the eligible losses.

In the pursuit of cheapness, some firms will obtain Bonds from dubious sources, so always check the source and pedigree of the Bond before allowing any contract to be signed. Legitimate Bonds can be secured from a firm's bank or insurance company or from a professional trade body to which they belong. Bonds can also be bought from firms that specifically provide that service.

Some organisations demand that Bonds use their own standard wording so as to ensure the level of 'cover' the client requires, and this can sometimes cause problems with the Bond supplier, who may want alternative wording. Such issues are normally resolved, albeit with the involvement of your legal department (because the Bond is a legal document), but it does mean the tender and award process may be delayed in the meantime.

Parent Company Guarantees

If a company has a parent company – in other words the bidder is a subsidiary of a larger Group – you should always ask for a Parent Company Guarantee (PCG). First of all they are free and secondly you benefit from the assurance that the parent company is willing to stand guarantee for their charge – which is always a good sign. If a company will *not* stand guarantee for their subsidiary, then you need to reconsider the whole deal.

As with a Bond, organisations will usually have their own form of Parent Company Guarantee with wording in their favour so some negotiation with the Parent Company may be required. It is rarely an issue, but PCGs can also come as on-demand or conditional, and again the conditional form should be used.

Novation

Novation, technically, is defined as the substitution of a new obligation for an old one. In our world this more specifically means the substituting a contract's provider with a new one. A Novation normally occurs in one of three instances, namely when:

a) A company goes into administration or liquidation part way through a contract and (often) the Administrator will sell the contract as an asset or
b) A company is bought out by another (or goes into a merger) part way through a contract
c) When you wish to transfer design liability to a constructor.

Scenario (a): A company goes into administration or liquidation part way through a contract.

Bonds and Guarantees will assist with the costs associated with a company going out of business part way through as contract but, as explained above, it will not pay for a new provider to complete the works. If another company wishes to buy the contract they may legally do so and this is called Novation, whereby the obligations inherent in the old contract are taken over by the new firm in a new one.

If a Novation 'crops up', your legal section should always be involved but there are some practical considerations from the procurement point of view that you need to take on board.

i) The contract can be bought provided you are in agreement. You need to be satisfied that the new company is suitable and to achieve this they need to go through the same selection process that the original firm went through, but with particular attention to their financial status (remember they are buying the contract).
ii) You should always seek a Bond or a PCG from the buying organisation
iii) They cannot compel you to agree the deal – although the Contract Administrator will be eager for you to do so as they will benefit from the proceeds.

iv) On this basis – do not be rushed.

v) The costs of picking up the loose ends left by the outgoing provider will need to be ascertained and processed as part of the claim against the original Bond or Guarantee, as will any legal costs associated with securing the Novation.

vi) If agreed, a new contract is drawn up using the same terms and conditions as the original as well as the same specification. The contract sum remains the same as well (i.e. they buy the existing contract).

Scenario (b): A company is bought out by another part way through a contract

Most of the points (i) to (vi) above apply, except that there is no longer an Administrator involved. You have to agree to the Novation or it cannot take place – it is a contract and requires the signatures of all parties.

What are the options?

The question arises, of course, 'what you do under circumstances (a) and (b) above if your conditions for a Novation are not met?' You may not like the proposed new company or they will not accept the standing terms and conditions. There may not *be* a buyer if your contractor goes out of business. The only real option you have is to re-tender the contract as it sits, plus a little bit for hand-over and pick-up from where the last provider finished.

However there are, as always, certain considerations that apply to complicate the mix.

Firstly, if the contract being Novated is currently with a company on a Framework, you cannot Novate the contract to a company not on the Framework. The contract was tendered specifically to firms on a Framework and this is in itself a contractual agreement. The new firm cannot join the Framework because the laws surrounding Frameworks do not allow it.

Secondly, challenges have, in the past, been raised by contractors when a contract is novated on the grounds that a new contract has been formed and no tender has taken place. This is an area that was open to much debate (a bit like aggregation and the Approved List) but the 2015 Regulations have come to the rescue and allow Novation under a range of circumstances, including when a company is taken over as a result of going into administration.

Our concern is primarily from the procurement aspect and this includes risk: if a Novation rears its ugly head, get legal on board straight away and seek advice on this. You need to be sure the incoming provider is able to perform the contract, and that their intentions are honourable – are they simply asset-strippers that are going to leave you in the lurch a little way down the line?

Thirdly, and this is also on the point of the contract's nature, if the contract is for an essential service, or works are at a critical stage, you may opt for Novation despite the fact that in many respects it is not the best option but it is the quickest route. The half-way house under these circumstances is to commission an emergency interim provision whilst an alternative is sought through more measured channels. Urgency (or emergency) will have a bearing on the approach you adopt.

Lastly, remember – Novation under these circumstances is never chosen by you. It is inflicted and often comes out of the blue, and you have to respond. Just keep calm and make measured decisions based on the facts, all the while guided by your legal support.

Scenario (c): Transferring a design liability to a constructor.

If you commission a designer (e.g. architect) for a project there will come a time when the constructor starts work and eventually the building will be finished. Problems can arise if something goes wrong during the construction stage or after completion: the high risk here is that there will be a conflict of liability in that the designer and the constructor may both contest to avoid liability – each blaming the other. To avoid this, it is common to make it clear at time of tender that, once the construction phase starts, the design contract or commission will be novated to the constructor and they will take on the liability. This makes it clean and simple. The constructors will, of course, want to vet the design and may well ask for a surety from the designer in the event that the design is the cause of any ensuing problem but, once the design is novated, only the contractor is responsible to you for the project completing successfully (and staying built). The only drawback is that – of course – the contractor, at time of tender, will build the cost of any risk this implies into his bid price but this will have been considered as part of the project plan and allowed-for in the costings.

Letters of Intent

These are the bête noire of lawyers – or they ought to be. Contract managers can sometimes be over-keen to use Letters of Intent to get the contractor on site sooner. Be wary of them (Letters of Intent, not contract managers): they are a letter asking (instructing) the contractor to start work now on the basis that the client intends to enter into a contract in the future, but no liability is accepted in regard to that intention. The idea is simple and logical, but like all simple, logical things, reality steps in and very often mucks it all up.

All too often, a reason crops up that prevents the contract going ahead as planned: for example, financial and political influences can all come to bear. If the contract does not go ahead, a Letter of Intent does not always resolve the ensuing issues regarding costs and liabilities. You are therefore advised to steer well clear of Letters of Intent unless you absolutely *have* to use one. If you do have to use one:

- Ensure legal draft it
- Make clear the scope of works covered by the Letter
- Limit the client's financial liability very specifically
- Use it on condition that the intended contract's terms and conditions apply.

Sealed Contracts

A normal contract, signed 'under hand', will be legally enforceable in line with the Statute of Limitations: that is, for up to six years from the last active date of the contract – this would, for us, normally be the final account settlement day.

Sometimes you want more security than that: if the contract is to build new houses, you will require a longer legal commitment than just six years and in this event you can have a contract sealed as a deed. This is a process executed by your legal team and will ensure the contract is enforceable for twelve years into the future.

Legal Precedents

As with any legislation, it is the ensuing court cases that determine what the law actually means and how it is interpreted. That is why law books are so fat – the regulations in themselves are quite thin (relatively speaking) but all the stuff that follows is what matters. The EU Regulations are just the same and you need to keep yourself up to date with court cases relating to EU Procurement Law in order to know how the tendering market is behaving (how many challenges are out there and in what areas) and how the law is being applied.

It would be a fool's errand to try and provide a list of cases for you to look at, although this book has been written with the latest cases in mind and so aims to reflect the impact of known case law to date[9]. Chapter 10 on the EU Standstill (Alcatel) period is a prime example of this and so are some of the issues discussed in Chapter 13 on sustainability. Chapter 7 pays great heed to recent case law when advising on evaluation criteria so, essentially, I have done about as much as I can within the scope of this book and the rest is up to you. You need to read the trade press and place yourself on e-bulletin distribution lists so that you get the latest information; legal firms often hold seminars specifically to update the audience

---

9. The reader is, however, referred to 'EU Procurement: Legal Precedents and their Impact' by the same author which does attempt to list and describe the most significant cases where precedents have been set.

on the latest rulings. Seek and ye shall find...

Part of my gene structure is anorak, and I tend to copy the most interesting or significant judgements and file them for future reference: I do not pretend that I can remember them all but they are very useful material for personal reference and for illustrating legal issues and associated risks when practising and when training other people. How you do it is up to you: just do something.

Learning more

Procurement is a discipline or profession where you will never stop learning. There is always something different to procure, new commercial developments to be aware of and new legal precedents that have to be borne in mind – such as those I have mentioned above. This is even more the case with the advent of the new Regulations: it is almost certain that they will be tweaked, new Guidance notes will be issued by the Cabinet Officer and legal cases will bring interpretations to bear. So, how do you keep up with it? How do you continue to learn?

Membership of the Chartered Institute of Procurement and Supply (CIPS) is essential if you are serious about procurement (see Links in the Appendices) and is a prime source of information and learning opportunities ranging from one-off evening events to structured courses leading to a Post-Graduate Diploma. Whilst there is not a wealth of books around on procurement, CIPS will have available the majority of what there is to be had.

In addition to this, there are many seminars and workshops held all round the country that will provide excellent tuition on different areas of the process. These are commercially sponsored events so I cannot name any specific provider here but anyone who is involved in procurement will be able to advise you where to look. Training is never wasted but be advised – some 'event' providers are better than others.

In addition, many law firms issue law updates and hold seminars on specific areas of procurement and contract law – some of them free – and again a little investigation will steer you towards them. To be good at procurement you need to keep up, so keep up.

Summary

- Procurement covers a multiplicity of areas – it is multi-disciplinary
- Every stage and every aspect is important – get it right
- Make sure you have the appropriate specialist skills and knowledge around you
- Distribute accurate and comprehensive documentation
- Investigate helpful tools and facilities such as Frameworks and Portals
- Aim to maximise the benefits of every procurement
- Never stop learning

17 // Contract Management

Does this have anything to do with procurement?

In a word (two words) – yes, definitely. I refer you to Chapter Six where I explained the procurement cycle which feeds lessons learned from one contract into the procurement of the next. The involvement of the contract manager in the procurement process is vital: you need to benefit from their knowledge and experience and have their input into the new contract, particularly on previous areas of weakness and the provision of adequate management tools.

The chapter on Risk should also have underlined to you the close relationship between the procurement process and the ensuing contract – the two cannot be separated.

This chapter is not a mini course on contract management – there are many books on that subject – but covers contract management from the procurement point of view.

Management Tools – Term Contracts

The nature of the tools will depend on the nature of the contract and most standard contract forms will provide them. Even if they do, you may still wish to change (amend) them to more specifically meet your own needs, or you may want to add or create them from scratch. In this context, I will look first at term contracts and consider projects afterwards.

There are a range of contract management tools you can build into a term contract at time of procurement and we will look at these now.

Break clauses and termination

The obvious tools are the 'heavy' brigade of termination and break clauses. The terms and conditions will include Termination Clauses that will enable you to end the contract in the event that the provider fails to deliver or commits some other awful sin (causing reputational harm to the client is often cited as one of these sins).

Break Clauses define stages at which the contract will naturally end, unless it is agreed by both parties that it should be continued. As an example, if you are tendering a seven-year contract, you can include Break Clauses that end the contract at the end of years three and five: if you are not happy with the service you can choose not to renew (or not to extend) it. The planned end-points will allow you to make alternative provision (e.g. another tender process) or, if there is insufficient time, you can continue the contract but not for the full remaining term (e.g. just for another 6 months until the new contract is in place).

This can be dangerous, though, because if a contract is not to be renewed you would have to advise the incumbent and you may well find the level of service will plummet. This would present other problems, as most of your other management tools become ineffective if there is no contract for the provider to prolong.

Default processes

Termination in the event of failure is a bit drastic if the misdemeanour is relatively minor, so there is often a default process for normal, operational failures. Examples of these would be failure to complete a task to schedule, missing appointments, poor workmanship, and so on. To deal with these types of incidents you would include a (say) three-stage escalation process. An example of this might be:

Stage 1 - Notification of the fault and a 'polite' request to make good within 36 hours.

Stage 2 – Notification of failure to respond to Stage 1; fault to be remedied within 24 hours or Stage three will be invoked.

Stage 3 - Getting a third party contractor (normally the back-up) to do the work at the cost of the incumbent, with client's administration fees added.

The administration fees must be justifiable – you cannot impose penalties – so make sure the level you set is reasonable and includes all aspects of securing the services of and paying the other provider.

The contract may link Defaults and Termination by setting a level of defaults that will trigger consideration of termination of the contract.

Performance indicators

All of the above will apply when things have gone wrong, but contract management should be about making it work, not waiting for it to fail. To this end, performance indicators serve to monitor contractor performance and not only trigger remedies to potential failure but also serve to recognise good performance – an equally important aspect of successful contract management.

Performance indicators (often referred to as KPIs – Key Performance Indicators) need to be chosen carefully. You can make them up or use industry standards but try and avoid some common traps:

- Do not try to impress by having loads of indicators. More data means more work.
- Do not ask for information that you cannot measure or cannot measure easily.
- Do not ask for data that you will not use – ensure it is all relevant.
- You can include administration tasks in the range – for example, submitting the monitoring data by the due date can itself be a KPI.
- Do not lay down remedial measures that may transpire to be too severe or unworkable – do not paint yourself into a contract management corner

The first bullet advises against inadvertently creating too much work for the contract management team. If you do not have an IT system that can provide the required KPI data, you should make it a requirement that the contractor provides the information by a given date: "...data to be provided by the 10th day of each calendar month...", for example.

KPIs are essentially targets, so adhering to *SMART*[1] criteria is a good guide in the first instance. They should be easy to measure and collect, so avoid obscure assessments: KPIs are meant to drive good customer service so why have measurements that your average customer could not understand?

Only ask for data or measurements that you are going to use to manage performance – do not let contract monitoring become an exercise in statistics simply because you can.

KPIs do not all have to deal with the operation of a contract on site. If you have experienced tardy administration in a contract in the past, include some of your contract support requirements in your performance assessment regime.

It can be a temptation, when drawing up remedial actions, to be over-severe. Sometimes this is due to failing to understand the impact of the proposed remedy and sometimes it is because you have got the target wrong: if a KPI turns out to be harder to achieve than you intended, relatively harsh remedies may have to be applied as a routine rather than as the one-off you intended.

It is possible to score KPIs, and these scores can be linked to other management tools such as the requirement of contractor action plans to address persistently poor performance or the consideration of enforcing Break or Termination Clauses.

I have in the past linked KPI scores with a cash bonus system – which becomes either a stick or a carrot depending directly on how the contractor performs. This can be very effective but can also be difficult to get the principle authorised by Council Members or Management Boards.

Contract progress meetings
Effective communication with a contractor is essential for good contract performance. The procurement process can lay down requirements in this regard, specifying what meetings should take place and when – from on-site liaison to formal progress meetings. The latter should be at least monthly for a normal term contract

---

1. SMART – Smart, Measurable, Achievable, Realistic and Time-bound.

(less 'busy' contracts may cope with quarterly meetings) and the contract should specify the minimum attendee list and set a standard agenda.

The standard agenda should include all the usual stuff:

- Introductions
- Apologies
- Performance
  - Outstanding orders
  - Late works
  - Default notices
- Complaints
- Health & Safety
- Finance / budget / Payments
- Risk Register
- Contractor's issues
- AOB
- Date and venue of next meeting

This is not _the_ Agenda, it is _an_ Agenda – you can make it what you like and what is suitable for the contract. You would, however, be wise to include an item that gives the contractor a slot to moan about the client and before AOB there may be a few more regular items or specific matters that require discussion at that particular meeting.

Training and guidance

Once you have tendered a contract, no-one will know more about it than you. For the contract to be managed effectively you will need to impart your knowledge and there are two very good ways of doing it. Firstly, I suggest you create a _Contract Guidance Manual._

A good Contract Guidance Manual will be the contract's handbook and have all the information essential for its effective management. It will hold basic information such as names of the providers and their contact details, including out-of-hours numbers, and information on any lots or packages, and so on. The manual will then go on to explain how this contract is different from its predecessor (remember – we are talking about term contracts) and point out key elements of the specification. The Manual will explain what management tools have been written into the contract and how they should be applied, and list all the performance indicators. It will have an Appendix with templates for all the documents relating to the contract, including the Meetings Agenda, Default Notices, etc. All the way through it will cross-reference specific clauses in the actual contract so that, if needs be, they can be referred-to or cited in any correspondence.

This sounds a major undertaking but it is not. Most of the information you will already have to hand because you wrote it, so most of the work will be to collate that information and translate any contract-speak into plain English for your colleagues to understand. Simple rule – keep it simple.

Once you have the manual in place, the second way to impart your knowledge is through training. In a two-hour session you will be able to explain all the intricacies of the new contract and equip colleagues to pick up the baton and manage it effectively.

There is a twist to this – it is a good idea to train the incoming contractors as well. Remember – the people who tendered for your contract will rarely be the ones who will run it (just like your lot) so to spend a couple of hours training the contractor's managers and supervisors will be of benefit to all parties and will hopefully be the start of an effective collaboration over the life of the contract. Try it.

Management Tools – Projects

The general principles of what has been said about term contracts will apply to projects, but with a few twists. Firstly, a major works or new build contract will be on site for a period of months instead of years and so the client/provider relationship is different. The contract management structure is also different, involving a Clerk of Works, a QS and a Surveyor in standard roles that are clearly defined by the contract.

Whilst you will build in some management tools, Standard Form contract clauses will be less prone to amendment than those in a term contract because the needs will tend to be less bespoke. The contract will have laid down a timetable and this will immediately establish the first KPI – when is the job to be finished by? The other KPIs will nearly always be selected from a range of industry standards that all major works contractors know and understand. There is little need to re-invent the wheel.

The meeting regime can be defined and again monthly is good, but fortnightly may be required if the project is likely to advance at a fast rate or there are particularly complex issues. The meetings Agenda must be fit for purpose and may be based on but will not be identical to the suggested format for a term contract, above.

Of course, Break and Termination Clauses are less relevant – if a contract is for six months' external decorations works, termination in the event of poor performance is not entirely feasible. You *can* terminate the contract, and it has been done, but the issues that arise from early Termination are really problematic – there is no back-up provider, the contractor may well just walk off site and leave health and safety and security issues abounding; there will be issues around the settlement of outstanding accounts, claims for damages to consider and so on. Termination really should be the absolute, absolute last resort, as a reality or as a threat.

Generally, an extreme contract management situation is less likely to occur. It is a short contract and generally more straightforward than a term contract to fulfil (term contracts can be very difficult 'on the ground'). A contractor will want to get off site and get paid as quickly as possible (maximising profits) but will not get away quickly if the standard of work is poor (when it will not be accepted by the client) or if they run behind schedule (obviously). For this same reason, escalating default processes are not required and all the standard contract forms provide for incidents such as tasks either not completed or completed late. One way standard forms provide for such eventualities is in the facilitation of LADs.

LADs

If a works contract runs late or fails in any area, the client can claim the costs associated with these misdemeanours. These claims are for tangible damages and losses and they are called liquidated and ascertained damages because they must be evaluated in terms of cost or cash (liquidated), proven as factual (ascertained) and be proper losses (damages).

LADs are a useful contract management tool as they help keep a contractor focussed on doing a proper job on time, but they must be properly applied. Firstly, they must not be – and not be used as – penalties. Hence the ascertained bit above – they must compensate the aggrieved party for actual losses in accordance with the criteria laid down in the tender documents. That is the second thing – you have to make it clear in the specification and T&Cs[2] when they can be applied and how the amounts are to be calculated.

An example of a valid claim would be the loss of rental income if premises are not ready for occupation by the contracted time, in which case the LADs would reflect the loss of rent on a weekly or daily rate basis.

Such losses cannot be claimed against a Bond but have to be deducted from monies owed to the contractor.

Extensions of time

If the contract does run late, as well as imposing LADs the client can refuse to grant an extension of time (EOT). An extension of time recognises that causes for a delay are out of the contractor's control (as defined within the terms of the contract) and so the costs associated with this will be met by the client. Such costs are normally associated with the preliminaries (the 'prelims') and comprise the costs of site compound, scaffolding hire, and so on. If the client considers the contractor to be liable for the delay, an EOT will not be granted and the contractor will have to bear all the associated costs as well as suffer any LADs that might be levied.

---

2. T&Cs – Contract terms and conditions

Resident consultation

A key area of difference in project contract management is the approach towards residents, where the consultation process already spoken of (Chapter 14) has given residents a seat in the cockpit – the project team meetings, giving them direct input into the way the contract is managed. Similarly, the need to access people's homes, a Resident Liaison Officer and a complaints book on site all serve to raise the profile of resident issues during the life of the contract, and contract management needs to take this aspect on board.

One final point

Lastly, a point for the contract manager: any contract will run better if you can develop a good working relationship with the contractor. Avoid being adversarial and always try to understand the problems the supplier may be having in fulfilling your requirements. This does not mean that every contract should be run as a partnering contract, nor that you give the contractor excessive leeway. What is does mean is that, although you do need to *manage* the contract, some give and take will (nearly) always be reciprocated, and if you give that extra bit you will generally get it back when you need it. If you find this 'positive' approach does not work you may have to change, but do not, from the outset, assume that managing a contract is always a battle of wits between you and 'them'.

Procurement's role

Everything above is about managing a contract, but all of it requires building in at time of procurement, for that is where the foundations for good and effective contract management are laid.

Summary

- Carefully plan contract management at the procurement stage
- Involve the contract's future manager as much as possible
- Communication is a vital contract management tool

- Contract Guide Notes and associated training will improve contract management beyond belief
- Use a range of complementary contract management tools
- KPIs must be reasonable and relevant
- Measures for covering losses are available, but you cannot impose penalties

APPENDICES

APPENDIX 1 – CORPORATE INFLUENCES ON TIMESCALES – AREAS OF DELAY

Stage	Comment	Time to allow
Processes:		
Approvals of: Procurement Strategy Award Recommendation	Approved in accordance with Scheme of Delegation (SoD). Timescale will be short if Head of Department, long if more Senior body such as Chief Executive, Mayor, Board or Cabinet.	Allow from 3 days to 4 weeks for each, depending on SoD, plus time required to write the Report and secure necessary concurrents/ comments.
Forward Plan	Constitutionally advises public of upcoming key decisions to allow them time for input. If it is not on the Plan, it cannot be 'heard'. Confirm your organisation's definition of a 'Key Decision'.	Check the Constitution – decision may require 120 days' notice or more.
Call-in period	High-level decisions will have a statutory 'call-in' period, giving time for a decision to be questioned.	Normally 5 days after the decision is published. Publication is normally via minutes of the relevant meeting and these can take 1 week (or more) to become officially 'published'. In all, allow at least 10 working days.
Scrutiny	Where a decision gets 'called in' – see above. The decision is subjected to 'trial' by a group of persons – normally Members - where the recommendation has to be justified. Not a good experience.	No defined timescale. Allow contingency of – say – two weeks in case it happens.

Gateway Process for any Report	Gateway Processes will normally apply to major decisions only.	Up to 6 weeks, plus time for all concurrents (allow up to four weeks*) plus time to write the Report. *Depends on number of tiers to be scaled and level of co-operation. In reality – I have known this process to take in excess of six weeks. If you don't know how long – ask. Procurement may require Gateway Reports at 2 stages – Pre-procurement and Award – but there can be more.
Consultation: Members	Key Cabinet or Board Members will probably have to be consulted, especially on high-profile tenders.	A few days should provide for Member consultation – depending on their availability – which will need to be checked. They may insist on particular days of the week – which may cause over a week's turn-round.
Stakeholders' or the public.	Stakeholders' consultation may be required for the report – especially on major works schemes such as block renovation.	Public consultation will normally require a meeting which, scheduled in advance, should add no time to the process, but maybe allow 3 days if you feel it necessary.
Leaseholders (in line with the Commonhold and Leasehold Reform Act, 2002)	Leaseholder consultation is a statutory requirement where Service Charges will be affected or contributions towards capital expenditure will be due. A minimum consultation period of two lots of 30 days is required by law, but extra should be allowed to deal with late queries. See Chapter 14.	Allow a minimum of 31 days and up to 42 days for each consultation stage.

APPENDIX 2 – PROCUREMENT TIMETABLE: EU Restricted procedure (note: this is a sample)

PROCUREMENT TIMETABLE:	CONTRACT NAME / TITLE: xxx		
Contract Value: £11m pa	Contract Period: 3 yrs + 2		Contract Start Date: 24th March 2015
Stage/task	Timescales		
	Cal. Days	Start	End
Pre-QS start – latest date	1	01 January 2014	01 January 2014
Agree Packaging & Strategy	3	02 January 2014	04 January 2014
Forward Plan – public and Gateway Board	120	09 February 2013	08 June 2013
Commission LH & Tnt Reps, stakeholders	5	05 January 2014	09 January 2014
Gateway 1 Report – Prepare	20	20 May 2013	08 June 2013
Gateway 1 – Process	30	09 June 2013	08 July 2013
Call-in Period	5	09 July 2013	13 July 2013
Notice of Intention – preparation	14	14 July 2013	27 July 2013
Draft Specification & Consultation	30	05 January 2014	03 February 2014
L/Holders Notice of Intention – Process	35	28 July 2013	31 August 2013
Forward Plan – Gateway 2 (on or before)	120	29 June 2014	26 October 2014
Prepare OJEU & trade adverts (book issue)	7	24 August 2013	30 August 2013
Preparation of Supplementary PQQ	10	13 August 2013	22 August 2013
Detailed & Final specification	40	04 February 2014	15 March 2014
Contingency Period	1	16 March 2014	16 March 2014
OJEU and Trade adverts	2	17 March 2014	18 March 2014
PQQ Period	30	18 March 2014	16 April 2014
PQQ evaluation and shortlisting	28	17 April 2014	14 May 2014
Contingency Period	10	15 May 2014	24 May 2014
Preparation of tender documentation	14	25 May 2014	07 June 2014
Collation of tender documentation/ bidders	7	08 June 2014	14 June 2014
Issue of tenders	1	15 June 2014	15 June 2014
Tender Period	30	16 June 2014	15 July 2014
Opening of Tenders	0	16 July 2014	15 July 2014
Tender Evaluation and 2nd stage appraisals	30	16 July 2014	14 August 2014
Contingency period	7	15 August 2014	21 August 2014
Invite to PTEs issued and lead-in time	10	22 August 2014	31 August 2014
Post-tender interviews	5	01 September 2014	05 September 2014
Gateway Board provision - prepare Gateway 2	14	06 September 2014	19 September 2014
L/Holders - Notice of Proposal – finalise	2	20 September 2014	21 September 2014
Notice of Proposal – Consultation period	35	22 September 2014	26 October 2014
Gateway 2 - Executive Approval	35	27 October 2014	30 November 2014
Call-in Period	7	01 December 2014	07 December 2014
Alcatel / standstill period.	12	08 December 2014	19 December 2014
Contract Award – pre-contract meeting / s	5	20 December 2014	24 December 2014
Lead-in time including TUPE	90	25 December 2014	24 March 2015
Start on site	1	25th March 2015	

Notes on this timetable:

This timetable shows 15 months start to finish including *all* preparatory work such as writing specifications. These processes may not all be required or take so long. The following comments also apply to this sample timetable:

- It starts from January 1st as an exemplar date
- It is based on a real tender exercise where all tasks had to be undertaken, including writing the specification, etc.
- It uses the Restricted Procedure with no timescale reductions employed
- It assumes a QS will be on board from the start of the process
- Assumes some timescales, e.g. 30 days for a Gateway process to complete
- Statutory Leasehold Consultation period is 30 days – 5 days added for contingency
- Assumes Post Tender Interviews will be held
- It includes two contingency periods
- It assumes 3 months' lead-in to accommodate TUPE transfer activities
- It allows 120 days for Forward Plan notification of Gateway Report hearings
- It does not allow for holidays (e.g. Christmas) or weekends
- In practice, once these off-days adjustments have been made, the timetable would be arranged in start-by-date order to give a chronological programme plan, to which a column can be added for RAG rating on progress against dates
- If this was a project, dates would be calculated forwards from approval to proceed with the project. In such a case...
- Some dates would still need to be calculated backwards, such as Forward Plan Notification, which has to be (say) 3 months *before* the actual Gateway hearing
- If it is a term contract, dates would be calculated *back* from the required contract start date to determine when the procurement needs to be commenced
- A spread sheet can have appropriate formulae set up to assist with making these calculations

APPENDIX 3 – STATUTORY REQUIREMENTS (Current at time of print)

Tendering authorities need to be aware of and observe the following, amongst others:

Procurement-related Law

Procurement is governed by a wide range of legislative and regulatory requirements, and procurement within Social Housing has a few extras as well. The following are some of the current applicable legislative controls:

- Public Contract Regulations 2015
- EC treaty of 1957 - introduced the principle of a single market and free movement of goods, services, workforces, and finance – Treaty of the Functioning of the European Union (TFEU)
- European Union Consolidated Directive on Procurement2004 – implemented into UK legislation by the Public Contracts Regulations 2006.
- The Public Procurement (Miscellaneous Amendments) Regulations 2011
- Government Procurement Agreement of the World Trade Organisation;
- Local Government Act 1972;
- Local Government (Contracts) Act 1997 (power to obtain goods, services and works);
- Local Government Act 2000 (power to promote well-being);
- Local Government Best Value (Exclusion of Non-Commercial Considerations) Order (Statutory Instrument 909/2001;
- DETR Circular 10/99.
- The Local Government Act 2000 (Constitutions) (England) Direction 2000.
- Construction (Design and Management) Regulations 2007 (CDM)
- Commonhold and Leasehold Reform Act 2002
- Public Services (Social Value) Act 2012

A search on the web will bring all these up, and some commentaries on them as well, so a little while spent browsing is highly recommended. Do not try and learn them – know where to find them.

Equalities

Public authorities have obligations and a general duty to provide and promote equality of treatment and opportunity. The Equality Act of 2010 replaced and encompassed The Race Relations (Amendment) Act 2000 and the Disability Discrimination Act 1995.

Health & Safety

There is a large volume of H&S legislation that needs to be considered, primarily the Health and Safety At Work Act, 1974. Do not be fazed – speak to your H&S Department. In the meantime, visit the Health and Safety Executive website at: http://www.hse.gov.uk/

Contract Management

The Housing Grants, Regeneration and Construction Act 1996 has recently been amended by the Local Democracy, Economic Development and Construction Act 2009. Known 'popularly' as The Construction Act, it lays down clear requirements of construction contracts, particular with regards to payment terms and practices.

Details on this and other Acts can be found at:
http://www.legislation.gov.uk/ukpga/2009/20/contents

APPENDIX 4 – SOME SOURCES OF GUIDANCE AND SUPPORT ON ISSUES OF SUSTAINABILITY

1. The OGC Archive has useful information and guidance on sustainability.
The OGC guidance can be found[1] at:

http://www.ogc.gov.uk/Introduction_to_Procurement_
sustainable_procurement.asp

2. WRAP – "WRAP (Waste & Resources Action Programme) helps local authorities [and other similar bodies], individuals and businesses to reduce waste and recycle more, making better use of resources and helping to tackle climate change". Their 'Halve Waste to Landfill' is worth looking at.

WRAP can be found at:

http://www.wrap.org.uk/

3. The Sustainable Procurement Cupboard: "The Cupboard was developed three years ago in response to procurement professionals request for a platform to share ways of implementing sustainable procurement practice in their organisations".

Sustainable Procurement Cupboard can be found at:

http://www.procurementcupboard.org/

4. SPIN: "SPIN is... dedicated to supporting local authorities in their efforts to procure in a sustainable manner, and provides a 'one-stop' website containing ...information relating to the sustainable procurement agenda."

---

1. Despite the OGC's move into the Cabinet Office, this web address still works

SPIN can be found at:
http://www.s-p-i-n.co.uk/

5. LGID: "LG Improvement and Development" (formerly the IDeA) supports improvement and innovation in local government, focusing on the issues that are important to councils and using tried and tested ways of working." They are able to provide support and advice on sustainability.

LGID (on sustainability) can be found at:

http://www.idea.gov.uk/idk/core/page.do?pageId=5246448

6. Constructing Excellence: "**Constructing Excellence** is the single organisation charged with driving the change agenda in construction. We exist to improve industry performance in order to produce a better built environment".

Constructing Excellence (on sustainability) can be found at:

http://www.constructingexcellence.org.uk/zones/
sustainabilityzone/commissions/msc.jsp

These are some of the key sources of information and advice. Many more are available.

APPENDIX 5 – SOME USEFUL WEBSITES

Item	Link
Cabinet Office:	http://www.cabinetoffice.gov.uk/
CIPS	http://www.cips.org/
CLARA 2002	http://www.legislation.gov.uk/ ukpga/2002/15/notes/contents
CompeteFor – see Glossary	https://www.competefor.com/
Crown Commercial Service – CCS	https://www.gov.uk/government/ organisations/crown-commercial-service e-mail at: localgovernment@ crowncommercial.gov.uk
EU General Guidance – Home Page	http://ec.europa.eu/index_en.htm
EU Regulations 2015	http://www.legislation.gov.uk/ uksi/2015/102/pdfs/uksi_20150102_ en.pdf
ISO	http://www.iso.org/iso/iso_catalogue.htm
Mytenders	http://www.mytenders.com/
OGC – Now moved into the Cabinet office. Archived OGC material can be found at:	http://webarchive.nationalarchives.gov. uk/20100503135839/http://www.ogc.gov. uk/index .asp
Public Services (Social Value) Act 2012	http://www.legislation.gov.uk/ uksi/2012/3173/pdfs/uksi_20123173_ en.pdf
SIMAP	http://simap.europa.eu/index_en.html
TED	http://ted.europa.eu/
UKAS – see Glossary	http://www.ukas.com/default.asp

APPENDIX 6 – PROCUREMENT TIMELINE - SAMPLE

ITEM / MILESTONE	1	2	3	4	5	6	7	8	9	10	11	12	13	14	15	16
BUSINESS CASE; PLANNING; REPORT; APPROVAL	▓															
SPECIFICATION		▓	▓													
CONSULTATION – STAKEHOLDERS – [NOT LEASEHOLDERS] - & INDUSTRY			▓	▓												
PREPARE PQQ				▓	▓											
WRITE TENDER SUBMISSION FORM					▓	▓										
WRITE TENDER EVALUATION MODEL						▓										
ISSUE OJEU ADVERT					▓											
PQQ PERIOD							▓									
EVALUATE PQQ								▓								
INVITATION TO TENDER (ITT)									▓							
TENDER PERIOD										▓						
EVALUATE TENDERS										▓						
PREPARE AWARD REPORT											▓					
AWARD REPORT; CALL-IN												▓				
ALCATEL / STANDSTILL													▓			
AWARD PROCESS													▓			
TUPE?														▓		
LEAD-IN															▓	
START ON SITE																▓

APPENDIX 7 – EU TIMESCALES AND WAYS OF REDUCING THEM

Procedure	PQQ – calendar days			Tender – calendar days		
	Normal	Using e-tendering	With PIN	Normal	Using e-tendering	With PIN
OPEN	N/a			35	30	15
RESTRICTED	30	30	15	30	25	10
COMPETITIVE WITH NEGOTIATED	30	30	15	30	25	10
INNOVATION PARTNERSHIP	30	30	15	30	25	10
COMPETITIVE DIALOGUE	30	N/a		N/a		
ACCELERATED OPEN	N/a			15		
ACCELERATED RESTRICTED	15	N/a		10	N/a	

Notes:

- You must *add* five days to minimum timescales if you are unable to issue *all* documents with the advertisement – even with the Restricted Procedure.

- You can use the PIN Notice concessions provided you go to the market at least 35 days *after* the PIN is published and *less* than a year after it is published.

- You *cannot* benefit from timescale reductions if you also use the PIN Notice as a call for competition

- For the Restricted and the Competitive with Dialogue Procedures, you can reduce to tender period to a minimum of 10 days provided all bidders agree. This means that if a bidder requires a minimum of 15 days (for example) and they all agree, you can reduce it to 15.

- Contract Award Notices must be published within 30 days of award, unless a quarterly Notice is being issued.

APPENDIX 8 – RIBA PROJECT STAGES
RIBA – Royal Institute of British Architects – Project Stages from 2013

ORIGINAL STAGES				NEW STAGES	
Preparation	A	Appraisal		1	Preparation
	B	Design Brief			
Design	C	Concept		2	Concept Design
	D	Design Development		3	Developed Design
	E	Technical Design		4	Technical Design
Pre-Construction	F	F1	Product Information		
		F2			
	G	Tender Documentation			
	H	Tender Action			
				5	Specialist Design
Construction	J	Mobilisation		6	Construction (Offsite and Onsite)
	K	Construction to Practical Completion			
Use	L	L1	Post Practical Completion	7	Use & Aftercare
		L2			
		L3			

APPENDIX 9 - GLOSSARY OF TERMS

Term	Description
Acceleration	The process of speeding up a procurement process in a manner defined and on grounds specified as acceptable within the EU Regulations.
Accreditation	The formal recognition by an accreditation authority of the technical and organisational competence of an organisation or firm to carry out a specific service in accordance to the standards and technical regulations laid down by the accrediting body.
Agreed Maximum Price	A price or rate for a contract or project, and usually the base for the sharing of any cost gains or losses against that price.
Alcatel	Name of the standstill period required by law for all EU procurements and mini-competitions held within a Framework.
Amendment (to a contract)	Change/s made to a (normally standard) Form of Contract to make it more fit for use by the tendering organisation.
Applicant	Term used for a firm who has submitted an Expression of Interest, i.e. applying to be selected to go on a tender list.
Approved List	A list of providers, all of whom have met corporate criteria and from which tender lists can be made up without recourse to any further pre-selection stage.
Articles (Contract)	Part of the preliminaries (qv), Articles set out facts about the contract's operation but do not lay down conditions or performance requirements.

Best Value and the "Four Cs"	The ethos of achieving a service that represents best value for money rather than simply the cheapest. The Four Cs represent C, C, C and C, which are seen as the four key elements of conducting a Best Value procurement. Now somewhat out of fashion or favour, this is still a principle with sound merits (see Introduction).
Best Value	A new type of Best Value, laying out 5 principles of procurement designed to steer authorities when working with voluntary and community groups and small businesses (see Chapter 15).
Bond	A Performance Bond underwrites the provider completing the contract and provides financial cover from any losses incurred in the event that they fail.
British Standard (BS)	Symbolised by the famous 'Kite Mark', this is a quality Accreditation. Using a 'BS N°' for each category, it lays down minimum standards for materials design and operational process. A BS standard can be used as a minimum specification for materials or a BS Accreditation can be demanded of a bidder as proof of good process. Often now superseded by the European ISO Accreditation (qv).
Building Surveyor	Specialist in the construction and fabric of a building who can analyse faults and draw up specifications.
Business Questionnaire	The equivalent of a Pre-Qualification Questionnaire but used in an Open Procedure, so not properly part of a pre-selection process.
Buying Solutions	A Government-sponsored body that tenders and operates framework agreements for supply of goods and consultancies to local authorities. Used to be part of the OGC, but is not any more - now in the Cabinet Office.
Call In	After a Key Decision is made, the decision may 'Called in' for Scrutiny or examination by a panel of peers as a way of ensuring good decision-making by Members and Council Officers. Normally there is a five-day (or so) standstill period after the decision is made public to allow call-in to take place.

CCT	Compulsory Competitive Tendering – a government initiative of the 1980s designed to improve financial efficiency in Local Authorities whereby everycouncil service had to be provided by an organisation who had won the contract to do so through a lowest-price tender.
Certification	The procedure by which a third party assures that a product, process, system or person conforms to specified requirements.
Challenge	An objection to a procurement process, normally but not necessarily in the first instance through a legal process, on grounds of contravention of the Regulations. Outcomes are often subject to or create legal precedents (qv).
CHAS	Contractors Health & Safety Assessment Scheme – an organisation "dedicated to completing health and safety pre-qualification assessments" a tendering organisation can use if it wishes.
CLARA	Commonhold and Leasehold Reform Act.
Clarification	Explanation sought or offered regarding the meaning or content of a tender or qualification package.
CLG	The Government's Communities and Local Government Department which is "responsible for maintaining and developing a framework for local government finance". Much of what local authorities do is steered by the CLG. It used to be called the Office of the Deputy Prime Minister – ODPM.
Client	To you: The person, persons or Business Unit for whom you are carrying out the tender process. Does not mean residents, who are referred to as customers. To the bidders and providers: The body that is responsible for payment for the services: may not be the end-user (see 'Customer').

CompeteFor	CompeteFor is a free service that enables businesses to compete for contract opportunities linked to major public and private sector buying organisations. (See Appendix 5).
Competitive Dialogue	A procurement route designed for complex projects that uses stages of dialogue with multiple bidders to arrive at the most appropriate solution via a final-stage tender evaluated against MEAT criteria.
Consortium	A link-up of firms to cover a range of disciplines in order to submit a bid. It is nota legal merger of the parties.
Contract Notice	An OJEU advert for a tender.
Contract Specific Questionnaire	Qualification Questionnaire relating to the contract being tendered; the same as a Technical or Supplementary Questionnaire.
Contract Standing Orders	Internal regulatory regime that lays down corporate processes for tendering and managing contracts.
CORGI	Confederation of Registered Gas Installers – regulatory body and precursor of Gas Safe (qv). Once a compulsory requirement for all tradesmen working on gas, this is now a defunct accreditation.
Corporate Questionnaire	Questionnaire used for selecting firms suitable to bid on the basis of compliance with corporate or organisation requirements pertaining to financial probity, health and safety practice, insurances, compliance with equality requirements and general business integrity. Notrelated to any specific contract, there are overarching minimum requirements for all suppliers.
COSHH	Control of Substances Hazardous to Health Regulations 2002
Crown Commercial Service – CCS	The Commercial Arm of the Cabinet Office and 'front line' for issuing Government procurement advice and guidance
CPV - Common Procurement Vocabulary Codes	Numerical codes used in OJEU adverts (or Contract Notices) to define the goods or services required of the bidders.
Customer	The end-user – normally residents.

'Dear John' letter	A letter of regret explaining that an applicant or bidder has not been successful at a stage in a procurement process. If issued in response to a full EU tender at tender stage, this letter will be in the form of a formal Alcatel letter.
Engrossment	The compiling of all documents required to make up the complete contract package.
E-tendering	The processing of a tender exercise through electronic (on-line) means.
EU	European Union
Exemption	See Waiver
Expression of Interest	A process whereby firms request a PQQ and complete and submit it for evaluation, thus expressing an interest in tendering for the contract opportunity.
Extension	A change to a contract in terms of its period, length or time. Granting an extension of time will often commit the employing authority to accepting any resulting cost increases.
(Standard) Form of Contract	An off-the-shelf contract as defined by the body that drew it up (e.g. JCT). Does not refer to the typeor natureof the services provided through it. Amendments are often made to Standard Forms to tailor them to the needs of the organisation issuing them.
Forward Plan	A list of upcoming Key Decisions to be made by a Local Authority and its Council Members so that the public may have notice of it and attend or have input. Often, dates have to published up to 3 months ahead of the date of decision (depending on the Council's Constitution).
Four Cs	See 'Best Value'
Framework and Framework Agreement	A Framework is a competitively tendered set-up of companies, all appointed so as to be able to provide the services for which the Framework was established. Selection of a provider on any one occasion can be by one of a variety of methods, specified within the Framework Agreement or Contract.

Gas Safe	Regulatory body for gas engineers. Accreditation is compulsory. This organisation replaced 'CORGI'.
Gateway Process	An hierarchical process of decision making that secures review and approval at the appropriate level at key stages in the procurement process so that the final decision-maker is fully informed (and assured) by the time the decision gets to be made. A process of risk management.
GC Works Contract	Suite of Contract Forms originally drawn up by the Government to commission the rebuilding of housing after the war.
HEFCE	Higher Education Funding Council for England – a body that distributes public money for teaching and research to universities and colleges.
Impact Assessment	An assessment of the effect or consequences of a project, process or event, normally on a defined group, with a view to steering the process and guiding the creation of mitigations or contingencies to deal with any adverse effects.
Input specification	Specification whereby the operational side is dictated by the 'how' and the 'when' in clear detail.
Invitation to Tender – ITT	The ITT invites a firm to tender, sets out the key criteria a bidder needs to meet, and explains how to submit a tender.
ISO	International Organisation for Standardisation. A body that accredits products as compliant with specification standards.
JCT	Joint Contracts Tribunal – a body that produces a range of standard forms of contract in collaboration with various representative organisations of the construction industry.
Key Decision	A 'major' decision as defined by the organisation. In procurement terms, normal criteria might comprise the value of the contract (e.g. over the EU threshold) or the extent of its impact (e.g. two or more Wards of a borough).

KPI – Key performance Indicator(s)	Standards by which a company's performance is judged. May be set as targets, and should be 'SMART'(qv). May be based on industry standard measures or can be set up 'locally' by the client body.
LADs – Liquidated and Ascertained Damages	Costs identified and proven to have been incurred as a result of a provider's error. Have to be specified at time of tender.
Legal Precedent	A legal decision or outcome in court, that provides an example or authority for judges deciding similar cases later.
Letter of Acceptance	Official acceptance of a company's tender. This is a legal commitment.
Letter of intent	Used to start a contract before a contract is completed (signed). Dangerous and unpopular with Local Authorities, they should always the financial liability of the client.
MTC or Measured Term Contract	A contract where works orders are issued over a period of years, within a specified geographical area; the work is measured and valued against a pre-priced Schedule of Rates
MEAT	Most Economically Advantageous Tender – tender evaluation based on a combination of quality and price, each scored in specific proportions.
Method statement	Statement explaining how a company will carry out a task or deal with an issue, submitted as part of a bid and becomes contractual upon award.
Moderation	The process whereby evaluators agree on a final score where they erstwhile had all scored differently.
NICEIC	The National Inspection Council of Electrical Installation Contractors. "nicky-eck" for short.
NEC Form of Contract	Suite of Contract Forms drawn up by the National Engineering Council

NJCC	National Joint Consultative Committee of Architects, Quantity Surveyors and Builders – now defunct and operating as the JCT.
Novation	The process of transferring a contract from one provider into the hands of another. Occurs when a company is bought out or when an existing provider goes into liquidation, in which case the proceeds of the sale of the contract go into the Administrator's 'pot' as part of the assets for distribution. Also used to 'move' a design contract over to an appointed constructor to create a Design and Build liability for the completion of a project.
OGC	Office of Government Commerce – OGC was theofficial source of procurement advice and good practice. Disbanded on 1stOctober 2011, procurement information is now available via the Cabinet office website at: http://www.cabinetoffice.gov.uk/ Data and information published by OGC can now be found on the National Archive website at: http://webarchive.nationalarchives.gov.uk/20100503135839/http://www.ogc.gov.uk/index.asp
OJEU	Official Journal of the European Union where all EU notices are published.
Open Procedure	A tender procedure whereby anyone can submit a bid and that bid must be evaluated if the company passes the 'suitability' questionnaire.
Output specification	Specification where the resultis specified but the 'how' is left to the discretion of the provider.
Parent Company Guarantee	A guarantee by a company for the performance of one of its subsidiaries in the event that the subsidiary fails to complete the contract.
Partnering	A method of managing a contract where the client and provider operate as equals, sharing risks, gains and losses. Nota legal partnership.

PAS 91	BSI PAS 91 is a publicly available specification (PAS) that sets out the content, format and use of questions that are widely applicable to prequalification for construction tendering.
PIN or Prior Information Notice	An advance warning, published in OJEU, of a contract intended to be advertised some time later. The purpose is to warn the market in advance and secure some idea of market interest in the proposal.
PFI	Private Finance Initiative - a project (usually very large) that uses private investment, repaid over the life of a long-term commitment.
PPC Form of Contract	Project Partnering Contract - Commercial Form for Partnering Project Contracts created by Trowers & Hamlins.
PPP	Public/Private Partnership - in essence the same as a PFI contract.
Precedent, Legal	A legal case where a judgement is reached that serves as a principle for succeeding cases. Often used by procurement practitioners to clarify an ambiguous point of law and so guide process and good practice. Of particular interest and importance as new laws, rules and amendments are introduced (i.e. to test and interpret the new requirements).
Preliminaries	Contract Preliminaries comprise recitals and articles which explain basic facts about the contract but do not lay down any conditions or deal with performance.
Pre-Qualification Questionnaire – PQQ	A questionnaire used to vet applicants and from them select bidders for a tender list as part of a procurement process.
Pre-Tender Estimate (PTE)	Estimate of the cost of a contract, carried out before tendering, normally to secure the budget and ensure the correct procurement route is chosen.
PRINCE 2	**PR**ojects **IN**a **C**ontracted **E**nvironment – an OGC Project Management Process (and qualification) designed to minimise risk.

Quality Assurance (QA)	Usually a recognised accreditation (such as ISO, BBA or BS) verifying adherence to strict quality control processes within the organisation.
QS	Quantity Surveyor – see below.
Quantity Surveyor (QS)	Qualified professional whose role is to manage the financial aspect of a contract. They will help with pre-tender cost estimates, evaluation models, and cost-certification during the life of a contract, and so on. A procurer's most valuable asset.
Quotation	A straightforward offer of a price to supply; no quality evaluation is involved and usually results in using the suppliers terms of business, not the client's.
Recitals (in a contract)	Part of the preliminaries (qv), Recitals specify facts about the contract and its documentation but do not define any operational conditions.
RIBA	The Royal Institute of British Architects
RICS	The Royal Institute of Chartered Surveyors.
Restricted procedure	Procurement Route that pre-selects bidders for a tender on the basis of suitability to work for the organisation and their capacity and capability to perform the contract
Risk	The chance and impact of an event happening that is prejudicial to the success of a project.
Scheme of Delegation	Specifies the level of (delegated) financial authority of staff or officers in an organisation's structure
Scrutiny	The process of examining a Council decision, normally undertaken by a panel or Scrutiny Committee through the process of Call-in (qv).
Services	For the application of EU thresholds, a Service can be defined as the provision of an ongoing service of a revenue nature that does not deliver a concrete asset (my definition). Building a house is works; maintaining it is a service. Consultancy services are all Services.

SIMAP	Web site providing access to the most important information about EU tendering. A mine of information.
SLA	Service Level Agreement; usually drawn up by different divisions within one organisation, an SLA lays down what services will be provided, under what conditions and in exchange for what level of remuneration. Essentially, a contract where one cannot legally be drawn up.
SMART	Accepted criteria for establishing performance targets. Acronym for S, M, A, Rand Tbased.
SME	Small and Medium-sized Enterprise, defined as an organisation employing no more than 250 staff
Stakeholder	Anyone who has a vested interest in the project or service under consideration. Can be an extensive list, so a Stakeholder Analysis could be used to identify key stakeholders who need to be involved at key stages.
Stakeholder Analysis	An exercise to identify all stakeholders in a project (such as a procurement), and assess their level of involvement in and potential impact upon the project.
Standstill Period	A period allowed after the proposed winner of a tender is made known to give other bidders time to consider the result and raise objections if they so choose. Also known as 'Alcatel'.
Supplementary Questionnaire	Qualification Questionnaire relating to the contract being tendered; the same as a Contract-Specific or Technical Questionnaire.
Surveyor	Usually means Building Surveyor – see above.
Technical Questionnaire	Qualification Questionnaire relating to the contract being tendered; the same as a Contract-Specific or Supplementary Questionnaire.
TED	Tenders Electronic Daily lists all OJEU Notices and archives them. Also the portal through which you can issue all of your OJEU Notices.

Term Contract	A contract set up for a given period of time to provide a stated range of services - e.g. maintenance and repairs. Often referred to as an MTC - MeasuredTerm Contract.
TPC Form of Contract	Term Partnering Contract – Commercial Form for Partnering Term Contracts created by Trowers & Hamlins.
TUPE	Transfer of Undertakings protection of Employment (Act). Refers to the law protecting employees when another firm takes over the contract on which they are employed. Known as 'Tew-pee'
UKAS	The United Kingdom Accreditation Service. Umbrella organisation for a range of accreditation bodies.
Value for Money – vfm	Defined by the HEFCE as obtaining the maximum benefit from the goods and services it both acquires and provides, within the resources available.
Variation	A change to a contract in terms of scope, price or quantities. Not time.
Waiver	Exemption from an internal regulatory requirement.

Index

19075140R00229

Printed in Great Britain
by Amazon